Institute of Mathematical Statistics
LECTURE NOTES–MONOGRAPH SERIES
Volume 12

An Introduction to Continuity, Extrema, and Related Topics for General Gaussian Processes

Robert J. Adler
Technion–Israel Institute of Technology

Institute of Mathematical Statistics
Hayward, California

Institute of Mathematical Statistics

Lecture Notes–Monograph Series

The production of the *IMS Lecture Notes–Monograph Series* is managed by the IMS Business Office: Jessica Utts, IMS Treasurer, and Jose L. Gonzalez, IMS Business Manager.

Library of Congress Catalog Card Number: 90-83130

International Standard Book Number 0-940600-17-X

Printed in the United States of America

ב"ה

To my family,

Nuclear and extended.

Happy is he who has inherited from his forbears,
Happy is he who has a niche in which to live.

(*Jerusalem Talmud*)

אשרי אדם שזכו לו אבותיו,
אשרי אדם שיש לו יתד במי להיתלות בה.

(ירושלמי, ברכות: פ"ד, ה"א)

PREFACE

... on what these notes *are* meant to be, and what they are *not* meant to be.

They *are* meant to be an introduction to what I call the "modern" theory of sample path properties of Gaussian processes, where by "modern" I mean a theory based on concepts such as entropy and majorising measures. They are directed at an audience that has a reasonable probability background, at the level of any of the standard texts (Billingsley, Breiman, Chung, etc.). It also helps if the reader already knows something about Gaussian processes, since the modern treatment is very general and thus rather abstract, and it is a substantial help to one's understanding to have some concrete examples to hang the theory on. To help the novice get a feel for what we are talking about, Chapter 1 has a goodly collection of examples.

The main point of the modern theory is that the geometric structure of the parameter space of a Gaussian process has very little to do with its basic sample path properties. Thus, rather than having one literature treating Gaussian processes on the real line, another for multiparameter processes, yet another for function indexed processes, etc., there should be a way of treating all these processes at once. That this is in fact the case was noted by Dudley in the late sixties, and his development of the notion of entropy was meant to provide the right tool to handle the general theory.

While the concept of entropy turned out to be very useful, and in the hands of Dudley and Fernique lead to the development of necessary and sufficient conditions for the sample path continuity of stationary Gaussian processes, the general, non-stationary case remained beyond its reach. This case was finally solved when, in 1987, Talagrand showed how to use the notion of majorising measures to fully characterise the continuity problem for general Gaussian processes.

All of this would have been a topic of interest only for specialists, had it not been for the fact that on his way to solving the continuity problem Talagrand also showed us how to use many of our old tools in more efficient ways than we had been doing in the past.

It was in response to my desire to understand Talagrand's message clearly that these notes started to take form. The necessary prompting came from Holger Rootzén and Georg Lindgren, who asked me to deliver, in (a very cold) February 1988 in Lund, a short series of lectures on Gaussian processes. On the basis that the best way to understand something is to try to explain it to someone else, I decided to lecture on Talagrand's results, and, after further prompting, wrote the notes up. They have grown considerably over the past two years, one revision being finished in (an even colder) February in Ottawa, under the gracious hospitality of Don Dawson and Co., and the last major revision during a much more pleasant summer in Israel.

I rather hope that what is now before you will provide not only a generally accessible introduction to majorising measures and their ilk, but also to the general theory of continuity, boundedness, and suprema distributions for Gaussian processes.

Nevertheless, what these notes are *not* meant to provide is an encyclopædic and overly scholarly treatment of Gaussian sample path properties. I have chosen material on the basis of what interests me, in the hope that this will make it easier to pass on my interest to the reader. The choice of subject matter and of type of proof is therefore highly subjective. If what you want is an encyclopædic treatment, then you don't need notes with the title *An Introduction to ....* What you need, unfortunately, is to turn to the journal literature. The key papers, in terms of the breadth of their coverage, are those of Dudley (1973), Fernique (1975), Jain and Marcus (1978), Talagrand (1987, 1988a), Samorodnitsky (1987a,b, 1990), and Adler and Brown (1986), with the brand new monograph of Pisier (1989) being the best source for the Gaussian process/Banach space interface. Piterbarg (1988), which I received only recently, has a wealth of information related to the extremal results of Chapter 5. A large part of the material in my notes comes from these sources, so you now know where to turn to for further detail. What is new in these notes is not so much the results, as the organisation into what I hope is a readable and logical whole. Very often this has lead to much easier and shorter proofs than those in the journal literature.

I am grateful to Rootzén and Lindgren for convincing me to put the effort into giving the lectures and preparing the original set of notes. It has been an educational experience for me at least. I hope that others will also find this material interesting and entertaining. Ron Pyke suggested publishing them in the IMS Lecture Notes-Monograh Series, and, as he was a decade ago during the writing of *The Geometry of Random Fields*, has been a quiet pillar of support and constant source of encouragement throughout. Many readers – too many to list, even if I could remember them all – of earlier versions have made comments and/or corrections that have affected the version you are now reading. Special mention, however, must go to two of these. Patrik Albin, initially during the birth of these notes in Lund, later during a phrenetic two month visit to the Technion, and finally at the other end of what must have been the all time hottest e-mail connection between Israel and Sweden, not only found a seemingly uncountable number of mistakes, he also corrected every correctable one he found. In particular, the proof in Chapter 4 of the lower bound for the majorising measure characterisation of sample path boundedness is his streamlined version of the original Talagrand argument, and many of the other proofs of that chapter also contain heavy contributions on his part. Raya Epstein Feldman led a group of fellow sufferers at Berkeley through Version 2 of the notes. Hopefully their pains, along with those of my colleagues at the Technion Probability seminar who partook of a similar experience, have lead to a better constructed final product. All the above have my heartfelt thanks.

Of course, there were other kinds of support as well, and my thanks go to the U.S. Air Force Office of Scientific Research (AFOSR 87-0298, 89-0261), the U.S.-Israel Binational Science Foundation (BSF 86-00285) and the Swedish Natural Sciences Research Council.

Finally, a technical note. Equations are numbered sequentially, beginning anew in each chapter. Theorems, lemmata, examples, etc. all share a separate, but common numbering system. The symbol ■ not only denotes the end of a proof, but has occasionally been used to denote the end of a discussion. The Exercises have a strange but self explanatory numbering system of their own that arose for historical reasons and that I was too lazy to change.

May, 1990
Haifa, Israel

CONTENTS

I. INTRODUCTION

1. The Basic Ideas.

It is my aim in these notes to treat two basic and closely related problems in the theory of Gaussian processes: the question of sample path continuity and the distribution of the supremum of a Gaussian process over a fixed set in its parameter space.

The basic approach will follow the modern attitude that the precise geometrical structure of the parameter space of a Gaussian process is of relatively little importance in determining sample path properties. Of more importance is establishing the "size" of the parameter space when measured in terms of a metric based on the covariance function of the process. There are various ways to measure this size, the best known, and easiest to handle, being via the notion of *metric entropy.* The most recent, and most powerful tool however is the notion of *mesure majorante*, or *majorising measure*, introduced by Preston (1972), (albeit not under this name), developed by Fernique in a series of papers starting with Fernique (1974), and recently shown by Talagrand (1987) to be the only tool currently available that provides necessary, as well as sufficient, conditions for the a.s. continuity of a general Gaussian process.

In between regular metric entropy and majorising measures lie some other notions, retaining the ease of application of metric entropy while giving stronger results. We shall consider two of these as well, once in studying continuity, and once while looking at suprema distributions.

Before we can start anything, however, we need to settle on terminology and notation.

$(\Omega, \mathcal{F}, P)$ will be a complete probability space that will remain fixed throughout the notes. A Gaussian random variable X with mean $\mu \in \Re$ and variance $\sigma^2 \in \Re_+$ is a real valued random variable such that for each $\lambda \in \Re$

$$Ee^{i\lambda X} = e^{i\mu\lambda - \frac{1}{2}\sigma^2\lambda^2},$$

or, equivalently, the law of X has density $\sigma^{-1}\phi((x-\mu)/\sigma)$, where

$$\phi(x) := (2\pi)^{-\frac{1}{2}} \exp(-x^2/2).$$

If $\mu = 0$ we call X *centered*, and if we also have $\sigma = 1$, then X is called *standard normal.* Since the transition from centered to non-centered Gaussians is via the easy addition of a constant we shall treat, almost exclusively, only centered Gaussian variables.

A (centered) Gaussian process is a family $\{X_t\}_{t\in T}$ of random variables, indexed by a parameter set T, such that each linear combination $\sum \alpha_t X_t$ is (centered) Gaussian. Whereas the use of the letters t and T seem to denote that the parameter is "time", the entire thrust of the modern theory of Gaussian processes is that T should be very general and neither theorems nor proofs should assume any particular structure for T. (In general we shall require only that T be a metric space, and that it be totally bounded – to be defined in a moment – in its metric.) For example, we shall consider examples in which T is a subset of $\Re^k$, $k > 1$ ("multiparameter time"), or the set of all squares within $[0,1]^k$, $k > 1$, or all convex sets in $[0,1]^k$, $k > 1$, etc. In none of these examples is T either one-dimensional, or ordered.

If X is a centered Gaussian process, then the above definition implies that the *covariance function* $R(s,t) = E(X_s X_t)$ on $T \times T$, which determines $E(\sum \alpha_t X_t)^2$, also determines the law of the entire process. It is easy to see that R must be positive semi-definite (i.e. $\sum_{s,t \in S} \alpha_s R(s,t) \alpha_t \geq 0$ for all $S \subset T$ and $\alpha_t \in \Re$). It is a basic result on Gaussian processes, following from the Kolmogorov extension theorem and the form of the multivariate Gaussian distribution, that given any positive semi-definite R on $T \times T$ there exists a centered Gaussian process on T with covariance function R.

Without further mention we shall always assume that T has a countable dense subset (which we shall often explicitly exhibit as part of an entropy calculation), and shall assume that X is a *separable* stochastic process. The importance of this assumption is that questions of continuity and boundedness of X over T then reduce to equivalent questions over the dense subset, and the fact that this is countable often makes for somewhat easier arguments.

This is all we need in the way of formalities. ■

By way of an introduction to what awaits us, consider the two problems of determining the a.s. continuity of X on T, and finding the distribution of $\sup_{t \in T} X_t$. Assume T is a metric space with metric τ, and consider

$$(1.1) \qquad \rho^2(u) = \rho_\tau^2(u) = \sup_{\tau(s,t) \leq u} E(X_s - X_t)^2 .$$

If X is to be continuous, then it is obvious that we must have $\rho_\tau(u) \to 0$ as $u \to 0$. In fact, this is a trivial necessary and sufficient condition for mean square continuity; c.f. Exercise 1.1. To obtain full, almost sure, sample function continuity we shall require more, and, in particular, we shall have to concentrate on the rate at which $\rho(u)$ tends to zero. The function ρ cannot, however, be of intrinsic importance, for we could change to an equivalent metric on T, without affecting the continuity or otherwise of X, but changing the specific behaviour of ρ.

Once we realise this fact, it is obvious that there is no good reason not to choose a metric on T that is equivalent to the basic metric, but easier to work with. An appropriate candidate is given via the covariance function of X as

$$(1.2) \qquad d(s,t) = [E(X_s - X_t)^2]^{\frac{1}{2}}$$

in a notation that will henceforth remain fixed. Actually, d is only a pseudo-metric, since $d(s,t) = 0$ does not necessarily imply that $s = t$. Nevertheless, we shall call d the *canonical metric* for T and/or X. Since the question of a.s. continuity makes little sense if we are not prepared to accept that X is at least mean square continuous, which, as we have just noted, occurs if and only if the covariance function is continuous over $T \times T$, we shall henceforth assume R continuous. In this case it is easy to see that if (T, τ) is compact, then τ-continuity and d-continuity are equivalent. Since this fact is crucial to our approach, we give a simple proof.

Firstly, note that since R is continuous, if X is d-continuous on T then it is trivially τ-continuous. The other direction is not immediate. The problem comes from points that are distant in the τ-metric, but close in the d-metric. (For example, if we take perhaps the simplest of all non-trivial Gaussian processes,

$$X(t) = A\cos(t - \phi), \qquad t \in [0, T],$$

with A a Rayleigh random variable with probability density $a\exp(-a^2/2)$, $a \geq 0$, and independent of ϕ uniform on $[0, 2\pi)$, then since X is periodic $d(s,t) = 0$ whenever $\tau(s,t) := |t - s|$ is a multiple of 2π. Hence τ-continuity of X does not immediately imply d-continuity.)

Suppose, however, that X *is* τ-continuous. For $\eta \geq 0$, let

$$A_\eta = \{(s,t) \in T \times T \colon d(s,t) \leq \eta\}.$$

Since R is continuous, this is a τ-closed subset of $T \times T$. Furthermore, $\bigcap_{\eta > 0} A_\eta = A_0$. Fix $\epsilon > 0$. Then, by the τ-compactness of T, there is a finite set $B \subset A_0$ (the number of whose elements will in general depend on ϵ) such that

$$\bigcup_{(s',t') \in B} \{(s,t) \in T \times T \colon \max\left(\tau(s,s'), \tau(t,t')\right) \leq \epsilon\}$$

covers A_η for some $\eta = \eta(\epsilon) > 0$. That is, whenever $(s,t) \in A_\eta$ there is a $(s', t') \in B$ with $\tau(s,s'), \tau(t,t') \leq \epsilon$. Note that

$$|X_t - X_s| \leq |X_s - X_{s'}| + |X_{s'} - X_{t'}| + |X_{t'} - X_t|.$$

Since $(s', t') \in B \subset A_0$, we have $X(s') = X(t')$ a.s. Thus

$$\sup_{d(s,t) \leq \eta(\epsilon)} |X_t - X_s| \leq 2 \sup_{\tau(s,t) \leq \eta} |X_t - X_s|,$$

so that the τ-continuity of X implies its d-continuity. ∎

Thus, at least for a while, we can concentrate on questions of continuity with respect to the canonical metric only, safe in the knowledge that if T is compact this is not a restriction.

It thus seems reasonable to now assume that continuity problems are going to be associated with the incremental covariance function $\rho_d(u)$ of (1.1). Life, however, is not quite so simple. Assume for the moment that T is continuous, in the sense that for all $u \leq \text{diam}(T)$, where

$$\text{diam}(T) = D(T) = \sup_{s,t \in T} d(s,t),$$

we have that there exists at least one pair of points $s, t \in T$ such that $d(s,t) = u$. Then it is immediate from the definition (1.1) that

$$\rho_d(u) = \sup_{d(s,t) \leq u} [E(X_s - X_t)^2]^{\frac{1}{2}} \equiv u.$$

Since this is independent of the covariance function it cannot be the correct measure with which to study continuity. The information which used to be in ρ has now somehow moved into the relationship between the canonical metric d and the parameter space T, and, in fact, is tied up in the "size" of T as measured in terms of d.

We denote by $N(\epsilon)$ the smallest number of closed d-balls of radius ϵ that cover T, and set $H(\epsilon) = \log N(\epsilon)$. The latter is called the metric entropy function for T (or X). We shall refer to any condition or result based on N or H as an entropy condition/result.

THROUGHOUT THESE NOTES WE SHALL ASSUME THAT THE FUNCTION $N(\epsilon)$ IS FINITE FOR ALL $\epsilon > 0$. THAT IS, WE ASSUME THAT T IS TOTALLY BOUNDED IN THE CANONICAL METRIC.

This is a basic assumption, without which many of the most elementary results following will not be true. However, it is so weak an assumption, that it never need worry us from the point of view of restricting applications of the general theory.

The first "general theory" result, is the following:

1.1 THEOREM. *A sufficient condition for the a.s. continuity of X on T is*

$$\int_0^\infty (\log N(\epsilon))^{\frac{1}{2}}\, d\epsilon < \infty. \tag{1.3}$$

(This theorem has a long history, with the first published versions due to Dudley (1967) and Sudakov (1969). Dudley (1973) has a full historical account of the various early versions with their various proofs, both correct and "in doubt".)

Much more is actually true within the framework of Theorem 1.1 than we have stated here. For example, the function $\omega(\delta) = \int_0^\delta (\log N_\epsilon)^{\frac{1}{2}}\, d\epsilon$, serves, up to a random multiplicative constant, as a modulus of continuity for X in terms of the canonical metric d. Consequently, investigation of the integral in (1.3) involves the study of a number of properties of X at once.

We shall see how to prove Theorem 1.1, along with a variety of related results, in Chapter 4. Note now, however, that when $\epsilon > \text{diam}(T)$ we have $N_\epsilon = 1$, so that the upper bound on the integral in (1.3) is really diam(T). Furthermore, since $N(\epsilon)$ is clearly a decreasing function of ϵ, the issue of finiteness in (1.3) is at zero.

We shall motivate the need for an entropy type condition with a variety of examples in the following two sections.

Now, however, let us turn briefly to the question of the behaviour of $\sup_T X_t$. To be honest, we should point out that in general finding the distribution of this supremum is an almost impossible problem. For example, precise formulae for

$$(1.4) \qquad P(\lambda) \;=\; P\{\sup_{t\in T} X_t \geq \lambda\}$$

are known for only six cases (i.e. six covariance functions) if X is required to be stationary, and in each one of these T is a finite interval in $\Re^1$. The best one can hope to get here are reasonable formulae for the asymptotic (as $\lambda \to \infty$) behaviour of $P(\lambda)$, and here there is a reasonably full theory.

It is easy to see (i.e. to guess) that the $\lambda \to \infty$ behaviour of $P(\lambda)$ will depend on two factors. The first, and most obvious, is any lack of homogeneity of X on T. If EX_t^2 is not constant over T, then it is reasonable to expect that the supremum will be achieved somewhere in the neighbourhood of the point (or points) of maximal variance. This is in fact the case.

The second factor will be the local smoothness, or lack thereof, of X. Again, it is reasonable to expect that the rougher X is near a point of maximal variance, the larger one can expect the supremum to be. Again this idea can be made rigorous and precise estimates and results proven. We shall return to this in detail in Chapter 5, after we have proven the rather surprising result that if $t_o \in T$ is such that

$$EX_{t_o}^2 \;=\; \sup_{t\in T} EX_t^2$$

then, at least for large λ, $P(\lambda)$ is not very different from $P\{X_{t_o} > \lambda\}$. Borell's inequality, Chapter 2, is the formal theorem, and in many ways is one of the building blocks of these notes. Some examples in Chapter 5 will show us that λ need not really be all that large for this result to be true. In fact, $\lambda \approx 1.5 \sup_{t\in T} (EX_t^2)^{1/2}$ is often large enough!

Now, however, the time has come to look at some examples. We start with a look at the continuity question for a specific family of centered Gaussian processes, with the aim of convincing you that you should, once and for all, forget that in the prehistory of Gaussian processes "t" was indeed time, and to show that it is neither T that is important, nor X, but rather the relationship between them as measured by the canonical metric d. Not surprisingly, the family that we shall take for our first class of examples is based on what is perhaps the most important of all stochastic processes, Gaussian or not – the Brownian motion. The family is composed of some of the extensions of this process to parameter spaces richer and more complex than the real line. This family of examples forms the content of the following section, and is required reading, in terms of providing motivation for the general theory.

2. The Brownian Family of Processes.

Let $(E, \mathcal{E}, \nu)$ be a σ-finite measure space. (Usually it will be a Euclidean space with the Borel σ-algebra and Lebesgue measure). A *Gaussian white noise* based on ν is a random set function W on the sets $A \in \mathcal{E}$ of finite ν-measure such that

(1.5) $\qquad W(A)$ is centered Gaussian and $EW^2(A) = \nu(A)$.

(1.6) $\qquad$ If $A \cap B = \phi$ then $W(A \cup B) = W(A) + W(B)$ a.s.

(1.7) $\qquad$ If $A \cap B = \phi$ then $W(A)$ and $W(B)$ are independent.

Property (1.6) encourages one to think of W as a random (signed) measure, although it is not generally σ-finite. We describe (1.7) by saying that W has *independent increments.*

To see that such a process exists, think of $\mathcal{E}$ as a parameter space T, with a covariance function on $\mathcal{E} \times \mathcal{E}$ given by

$$R_\nu(A, B) \;=\; EW(A)W(B) \;=\; \nu(A \cap B).$$

Since for any $A_i \in \mathcal{E}$ and $\alpha_i \in \Re$

$$\begin{aligned}\sum_{i,j} \alpha_i R_\nu(A_i, A_j)\alpha_j &= \sum_{i,j} \alpha_i \alpha_j \int_E I_{A_i}(x) I_{A_j}(x)\, \nu(dx) \\ &= \int_E \Big(\sum_i \alpha_i I_{A_i}(x)\Big)^2 \nu(dx) \\ &\geq 0,\end{aligned}$$

R_ν is positive semi-definite. Thus W exists as a centered Gaussian process on $\mathcal{E}$. ■

To see what the canonical metric d is in this case, note that

$$\begin{aligned} d^2(A,B) &= E(W(A)-W(B))^2 \\ &= \nu(A)+\nu(B)-2\nu(A\cap B) \\ &= \nu(AB^c \cup A^c B) \\ &= \nu(A\Delta B) \end{aligned}$$

where the symmetric difference $A\Delta B$ is made up of the points in either A or B, but not both.

We specialize to the case $E = \Re^k_+ = \{(t_1, ..., t_k) : t_i \geq 0\}$, $\mathcal{E}$ = the Borel σ-algebra on $\Re^k_+$ and $\nu = \lambda$ = Lebesgue measure. At the risk of occasional confusion we denote a point $(t_1, \ldots, t_k)$ in $\Re^k_+$ simply by t. Let $(a,b] \subset \Re^k$ be the "half open" k-dimensional interval $\Pi_{i=1}^k (a_i, b_i]$, (with open, closed, and half-open from above intervals written and defined similarly), and set $W_t = W((0,t])$. W_t is called the *Brownian sheet.* Via this definition and the above, W is the centered Gaussian process with covariance

$$EW_s W_t = (s_1 \wedge t_1) \times \cdots \times (s_k \wedge t_k).$$

When $k = 1$, W is the standard Brownian motion. When $k > 1$, then if we think of $W(A)$ as a measure, W_t is the corresponding "distribution function". It vanishes on the axes, and if we fix $k-1$ of the indices, it is a scaled Brownian motion in the remaining free variable. (c.f. Exercise 2.1.)

The Brownian sheet holds roughly the same place in the theory of multiparameter stochastic processes and multivariate statistics that the standard Brownian motion does in one dimension. It is a multi-parameter martingale (Cairoli and Walsh (1975), Wong and Zakai (1976)) and forms the basis of the multiparameter stochastic calculus. There is a nice review of its basic properties in Walsh (1986), where you can also discover its central rôle in the theory of stochastic partial differential equations, and in what sense it is valid to describe the derivative $\partial^k W(t_1, \ldots, t_k)/\partial t_1 \ldots \partial t_k$ as Gaussian white noise. Markovian properties, however, are somewhat missing, due to the fact that the parameter space is not totally ordered. (But see Section 5.5 for one way to get around this, and keep your eyes open for a forthcoming paper by John Walsh and Robert Dalang on this topic.) The lack of order in $\Re^k_+$ will not worry us, however, since from the Gaussian point of view we have already agreed to forgo using the special structure of specific parameter spaces.

The Brownian sheet arises in multivariate statistics in two main settings. If Z^k_+ is the k-dimensional integer lattice in $\Re^k_+$ and $\{X_i\}_{i\in Z^k_+}$ a collection of centered, i.i.d. random variables with $EX_i^2 = 1$, then it is not too hard to show that $n^{-k/2} \sum_{i\in n(t)} X_i$ converges weakly to W_t as $n \to \infty$, $t \in [0,1]^k$, where $n(t) = \{i \in Z^k_+ : 1 \leq i_j \leq [nt_j],\ j = 1, ..., k\}$. ($[x]$ is the "integer part" of x.) That is, W_t is basic to the k-dimensional functional central limit theorem.

Secondly, let $X_1, X_2, \ldots,$ be a sequence of i.i.d. random variables, with common uniform distribution on $[0,1]^k$. Then the empirical distribution function $F_n(t)$ of the X_i is defined as

$$(1.8) \qquad F_n(t) = \frac{\#\{i \le n : X_i \in [0,t]\}}{n}.$$

Similarly, the empirical measure is defined as

$$(1.9) \qquad \nu_n(A) = \frac{\#\{i \le n : X_i \in A\}}{n}.$$

Then, as $n \to \infty$, the normalized empirical distribution function $G_n(t) := \sqrt{n}\{F_n(t) - F(t)\}$ converges weakly to the so-called *pinned Brownian sheet*, $\mathring{W}$, a version of which can be obtained as

$$(1.10) \qquad \mathring{W}_t = W_t - |t| W_1, \qquad t \in [0,1]^k,$$

where $|t| = \prod_{i=1}^k t_i$. If we restrict A to a small enough (in terms of entropy) class of sets, then $n^{\frac{1}{2}}\{\nu_n(A) - \lambda(A)\}$ converges weakly to a process we shall call the *set indexed, pinned, Brownian sheet*, a version of which is given by

$$\mathring{W}(A) = W(A) - \lambda(A) W([0,1]^k).$$

(For details about the weak convergence, consult Dudley (1978a) and Pollard (1984). For more properties of the limit processes, be patient.)

Our first result is

1.2 PROPOSITION. *The Brownian sheet and pinned Brownian sheet are continuous on* $[0,1]^k$.

PROOF: Clearly, we need only prove the proposition for W. The continuity of $\mathring{W}$ then follows from that of W and (1.10).

We shall use the entropy condition of Theorem 1.1, for which we must calculate a bound for the minimal number of d-balls of radius ϵ needed to cover $[0,1]^k$.

Fix $t \in [0,1]^k$, and let $S(t,\delta)$ be the k-dimensional cube with lower left corner t and sides of length δ: i.e.

$$S(t,\delta) = \{s \in [0,1]^k : t_i \le s_i \le t_i + \delta,\ i = 1, \ldots, k\}.$$

Then it is easy to see that

$$\sup_{s \in S(t,\delta)} E(W_t - W_s)^2 = k\delta.$$

(Equality holds here if $t + \delta \in [0,1]^k$.) Hence,

$$\sup_{s \in S(t,\delta)} d(s,t) \le \sqrt{k\delta},$$

and so one way to cover $[0,1]^k$ with d-balls of radius ϵ is with the $[k(1+\epsilon^{-2})]^k$ balls of radius ϵ^2/k centered at the points t of the form $(i_1\epsilon^2/k, \dots, i_k\epsilon^2/k)$, $i_j = 0, 1, \dots, [k\epsilon^{-2}]$. It thus follows that $N(\epsilon) < k^k(1+\epsilon^{-2})^k$, and so the entropy integral $\int_0^\infty (\log N(\epsilon))^{\frac{1}{2}} d\epsilon$ is finite with room to spare. This completes the proof. ■

The natural question to ask now is whether or not continuity is preserved when we look at Gaussian white noise indexed by more general classes of sets than the positive orthants that gave us the Brownian sheet. The motivation behind this question is that the weak convergence referred to above in the context of empirical processes requires that the limit process be continuous, and so the very *raison d'être* of our interest in the Brownian sheet (at least from this viewpoint) hinges on this question. The reason to look at more general index sets lies in the fact that the power of Kolmogorov-Smirnov tests based on statistics of the type

$$\sup_{A \in \mathcal{A}} n^{-\frac{1}{2}} |\nu_n(A) - \lambda(A)| \tag{1.11}$$

is substantially increased by taking richer classes of sets $\mathcal{A}$ other than merely lower orthants, (Pyke (1984, 1985)).

We shall give a number of examples of interesting classes $\mathcal{A}$ in the following section, and later in these notes. Now, however, we shall consider an example of a parameterising class of sets which is too rich to allow continuity.

Introduce a partial order on $\Re^k$ by writing $s < (\le) t$ if $s_i < (\le) t_i$ for all $i = 1, .., k$. Then a set A in $\Re^k$ is called a *lower layer* if $s \le t$ and $t \in A$ implies $s \in A$. With some looseness of terminology, we shall refer to the Gaussian white noise based on Lebesgue measure on a class of sets $\mathcal{A}$ as the Brownian sheet on $\mathcal{A}$.

1.3 PROPOSITION. *The Brownian sheet on lower layers in $[0,1]^2$ is discontinuous and unbounded with probability one.*

PROOF: We start by constructing some examples of lower layers. Write a generic point in $[0,1]^2$ as (s,t) and let $T := T_{01}$ be the right triangle in which $s \le 1$ and $t \le 1 \le s + t$. Let C_{01} be the square where $\frac{1}{2} < s \le 1$ and $\frac{1}{2} \le t \le 1$.

For $n = 1, 2, \dots,$ and $j = 1, \dots, 2^n$, let T_{nj} be the right triangle defined by $s+t \ge 1$, $(j-1)2^{-n} \le s < j2^{-n}$, and $1 - j2^{-n} < t \le 1 - (j-1)2^{-n}$. Let C_{nj} be the square filling the upper right corner of T_{nj}, in which $(2j-1)2^{-(n+1)} \le s < j2^{-n}$ and $1 - (2j-1)2^{-(n+1)} \le t < 1 - (j-1)2^{-n}$.

structures of Figure 1.1, where each step comes from the horizontal and vertical sides of some T_{nj} with, perhaps, different n.

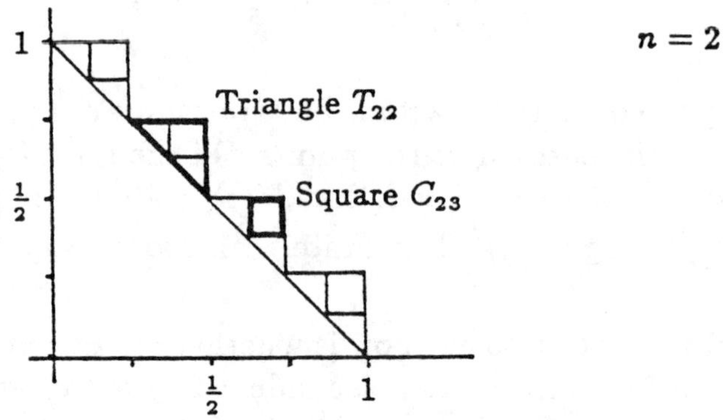

FIGURE 1.1. Construction of some lower layers.

Note that since the squares C_{nj} are disjoint for all n and j, the random variables $W(C_{nj})$ are independent. Also $\lambda(C_{nj}) = 4^{-(n+1)}$ for all n, j.

Let D be the negative diagonal $\{(s,t) \in [0,1]^2 : s+t = 1\}$, and $L_{nj} = D \cap T_{nj}$. For each $n \geq 1$, each point $p = (s,t) \in D$ belongs to exactly one such interval $L_{n,j(n,p)}$ for some unique $j(n,p)$.

For each $p \in D$ and $M < \infty$ the events

$$E_{np} := \{W(C_{n,j(n,p)}) > M2^{-(n+1)}\}$$

are independent for $n = 0, 1, 2, \ldots$, and have the same positive probability. (Since $W(C_{nj})/2^{-(n+1)}$ is standard normal for all n and j.) Thus, such an event occurs with probability one. Let $n(p) := n(p,\omega)$ be the least such n, defined and finite for almost all ω.

Since the events $E_{np}(\omega)$ are measurable jointly in p and ω, Fubini's theorem implies that, with probability one, for almost all $p \in D$ (with respect to Lebesgue measure on D) some E_{np} occurs, and $n(p) < \infty$. Let

$$\begin{aligned} V_\omega &= \bigcup_{p \in D} T_{n(p),j(n(p),p)}, \\ A_\omega &:= \{(s,t) : s+t \leq 1\} \cup V_\omega, \\ B_\omega &:= A_\omega \setminus \bigcup_{p \in D} C_{n(p),j(n(p),p)}. \end{aligned}$$

Then A_ω and B_ω are lower layers. Furthermore, almost all $p \in D$ belong to an interval of length $2^{\frac{1}{2}-n(p)}$ which is the hypothenuse of a triangle with the square $C_p = C_{n(p),j(n(p),p)}$ in its upper right corner, for which $2W(C_p) > M2^{-n(p)}$. Consequently,

$$W(A_\omega) - W(B_\omega) \geq \sum_p M2^{-n(p)}/2,$$

where the sum is over those $p \in D$ corresponding to distinct intervals $L_{n(p),j(n(p),p)}$. Since the union of the countably many such intervals is almost all of the diagonal, the sum of $\sum 2^{-n(p)}$ is precisely 1.

Hence $W(A_\omega) - W(B_\omega) \geq M/2$, implying that $\max\{|W(A_\omega)|, |W(B_\omega)|\} \geq M/4$. Sending $M \to \infty$ we see that W is unbounded and so, *a fortiori*, discontinuous with probability one over lower layers in $[0,1]^2$. ■

The above argument is due to Dudley (1978b), and a similar argument shows that W is unbounded over the convex subsets of $[0,1]^3$. Entropy arguments show that W is also unbounded over the convex subsets of $[0,1]^k$ for all $k \geq 4$, and (just to make sure that you don't confuse sample path properties with topological properties of the parameter space) that W *is* continuous over convex subsets of the unit square.

To see how entropy enters into this example, let us consider one further example, that is closely related to the above in terms of entropy, but is geometrically less appealing than lower layers.

Let γ be positive, and define the collection $\mathcal{A}_\gamma$ of subsets of $\Re^2$ as follows: Given $A_0 = [0,1]^2, A_1, \ldots, A_{n-1}$, let A_n be the closed rectangle whose left (vertical) side is the right side of A_{n-1}, (so that it has height 1) and with width $2^{n(1-\gamma)}$. Now divide each A_n into 2^n equal horizontal slices $A_{n1}, \ldots, A_{n2^n}$. The area of each A_{nk} is $2^{-n\gamma}$. The family $\mathcal{A}_\gamma$ is defined as the collection of all finite and countable unions of the sets $\{A_{nk}\}$.

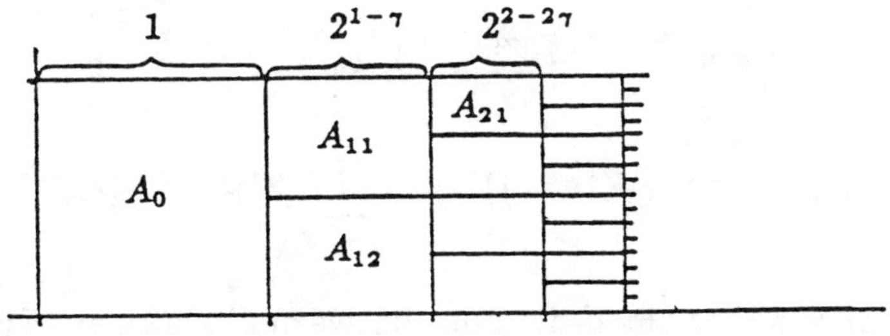

FIGURE 1.2. The partition used to create $\mathcal{A}_\gamma$.

We start by claiming that the entropy function $N(\epsilon)$ of $\mathcal{A}_\gamma$ satisfies, for all $\gamma > 1$

$$a_1 \exp(b_1 \epsilon^{-2/(\gamma-1)}) \leq N(\epsilon) \leq a_2 \exp(b_2 \epsilon^{-2/(\gamma-1)}), \tag{1.12}$$

for suitable constants a_i, b_i. It thus follows from Theorem 1.1 that since

$$\int_0 (\log N(\epsilon))^{\frac{1}{2}} \, d\epsilon \leq K \int_0 \epsilon^{-1/(\gamma-1)} \, d\epsilon,$$

the Brownian sheet will be continuous on $\mathcal{A}_\gamma$ as long as $1/(\gamma-1) < 1$, i.e. if $\gamma > 2$. We shall prove half of (1.12) in a moment.

for suitable constants a_i, b_i. It thus follows from Theorem 1.1 that since

$$\int_0 (\log N(\epsilon))^{\frac{1}{2}}\, d\epsilon \;\le\; K\int_0 \epsilon^{-1/(\gamma-1)}\, d\epsilon,$$

the Brownian sheet will be continuous on $\mathcal{A}_\gamma$ as long as $1/(\gamma-1) < 1$, i.e. if $\gamma > 2$. We shall prove half of (1.12) in a moment.

On the other hand, since for $1 < \gamma \le 2$ it follows from the lower bound in (1.12) that the entropy integral diverges, it would seem that in this case the Brownian sheet on $\mathcal{A}_\gamma$ is discontinuous. That this is in fact the case can be shown via an almost identical (but somewhat easier) proof to that used above to show unboundedness on lower layers. The details are left to Exercise 2.2.

The only case left undecided is $\gamma \le 1$. In this case $\bigcup_{n\ge 0} A_n = [0,\infty) \times [0,1]$, and so the collection of sets in $\mathcal{A}_\gamma$ is unbounded, and thus uninteresting.

We now establish the upper bound in (1.12), which is all we need to prove continuity. (The lower bound requires considerably more work and is derived in Samorodnitsky (1987a, 1990).)

Choose $\epsilon = 2^{-m}$, $m > 1$ and set $S_\gamma = 2^\gamma/(2^\gamma - 2) = \lambda(\bigcup_{n,k} A_{nk})$ (since $\gamma > 1$). Define

$$M \;=\; M(\epsilon) \;:=\; 1 + [(2m + \log_2 S_\gamma)/(\gamma - 1)].$$

Let C_M be the set $C_M = \bigcup_{n=M}^{\infty} \bigcup_{k=1}^{2^n} A_{nk}$. Then

$$E\big(W(C_M)\big)^2 \;=\; \sum_{n=M}^{\infty} 2^{n(1-\gamma)} \le 2^{-2m}.$$

Thus it follows that the d-distance between C_M and any subset of it composed of unions of A_{nk}, $n \ge M(\epsilon)$, $1 \le k \le 2^n$, is less than $2^{-m} = \epsilon$, and so all of these unions lie within a d-ball of radius ϵ of C_M.

We still need to find d-balls that cover all unions of A_{nk}, $n < M$, $1 \le k \le 2^n$. There are $1 + 2 + 2^2 + \cdots + 2^{M-1} = 2^M - 1$ such sets A_{nk}, and so $2^{2^M - 1}$ possible unions. Take each one of these as the center of a d-ball. Then

$$N(\epsilon) \;\le\; 2^{2^M} \le a_2 \exp(b_2\, \epsilon^{-2/(\gamma-1)})$$

for appropriate a_2, b_2. This is what we wanted to prove. ∎

To be certain that you understand this example, you should make certain that you understand why making γ larger makes $\mathcal{A}_\gamma$ "smaller". (After all, all $\mathcal{A}_\gamma$'s can be trivially mapped on to one another.)

The aim of the examples of this section was to convince you that the relationship between a Gaussian process and its parameter space is, as far as continuity and boundedness are concerned, an important subject. Note that all the different types of Brownian process that we saw here, whether they were simple Brownian motions defined on $\Re_+$ or complicated set indexed processes defined on a class of sets $\mathcal{A}$, could be handled by a common analysis, based on the covariance function and the canonical metric d. When the parameter space was more complex, so was the analysis; but *in principle* it remained the same. This would not have been so had we attempted to analyse these processes via extensions of the simple Markovian properties exhibited by Brownian motion on $\Re_+$ – these do not extend in any simple fashion to more general parameter spaces.

We shall now look at a number of other examples, some of which will seem at the outset to be qualitatively beyond the range of processes we have set out to consider. That they can be handled within the framework of real valued Gaussian processes on a totally bounded metric space is part of the justification of the general approach that we take. At this stage you should definitely work through the example of processes on $\Re^1$, in order to have a concrete example on which to hang the theory of the following chapters, and that on generalised random fields, in order to have some idea why there is a need to build a general theory that is independent of the structure of the parameter space. Beyond that, the choice of how far to continue is up to you. Some readers criticized early versions of these notes, because there were not enough examples to enable one to appreciate how the general theory could be applied. Other readers criticized later versions, because there were too many examples, and working through them they lost sight of the forest because of the trees. The choice as to how much you should read now is therefore up to you.

3. A Collection of Examples.

1. PROCESSES ON $\Re^1$: Returning from the generalisations of the previous section and the abstraction of the first to "simple" processes on the real line, it is natural to expect that conditions for continuity and boundedness will become simple to both state and prove, and there will be little, if any, need to talk about concepts such as entropy.

This expectation is both true and false. It turns out that avoiding the notion of entropy does not make it any easier to establish continuity theorems, and, indeed, reliance on the specific geometry of the parameter space often confounds the basic issues. On the other hand, the following important result is easy to state without specifically referring to any abstract notions. To state it, let X_t be a centered Gaussian process on a finite interval $[0, T]$, and define

$$p^2(u) = \sup_{|s-t| \le u} E|X_s - X_t|^2. \tag{1.13}$$

i.e. p is identical with the function ρ of (1.1), except that we are now using the usual Euclidean metric for the real line.

1.4 THEOREM. *If, for any $\delta > 0$,*

$$\int_0^{\delta} (-\log u)^{\frac{1}{2}}\, dp(u) \;<\; \infty, \tag{1.14}$$

then X is continuous on $[0,T]$ with probability one. Furthermore, a sufficient condition for (1.14) to hold is for some $0 < C < \infty$ and $\alpha, \eta > 0$

$$E|X_s - X_t|^2 \;\le\; \frac{C}{\big|\log|s-t|\big|^{1+\alpha}}, \tag{1.15}$$

for all s,t with $|s-t| < \eta$.

PROOF: The proof that (1.15) implies (1.14) (along with a similar, and somewhat easier condition to work with) is left to the exercises. Here we shall prove that the fact that (1.14) implies continuity is a trivial consequence of the entropy condition (1.3).

Note first that since $p(u)$ is obviously non-decreasing in u, the Riemann-Stieljes integral (1.14) is well defined, as is the inverse function $p^{-1}(u) := \sup\{t\colon p(t) \le u\}$.

Note also the obvious fact that, for each $\epsilon > 0$, the interval $[0,T]$ can be covered by $1 + [(2p^{-1}(\epsilon))^{-1}T]$ intervals, each of which has radius ϵ in the canonical metric d, (c.f. (1.2) and (1.13)). Thus,

$$\begin{aligned}
\int_0^{\infty} \big(\log N(\epsilon)\big)^{\frac{1}{2}}\, d\epsilon &= \int_0^{p(T/2)} \big(\log N(\epsilon)\big)^{\frac{1}{2}}\, d\epsilon \\
&\le \int_0^{p(T/2)} \Big(\log\big(1 + [T/(2p^{-1}(\epsilon))]\big)\Big)^{\frac{1}{2}}\, d\epsilon \\
&= \int_0^{T/2} \Big(\log\big(1 + T/(2u)\big)\Big)^{\frac{1}{2}}\, dp(u).
\end{aligned} \tag{1.16}$$

The question of the divergence of the last integral hinges on its behaviour in the neighbourhood of zero, and it is clear that (1.14) implies its convergence. Thus, by Theorem 1.1, we have the a.s. continuity of X. ■

Whereas the integral condition (1.14) is sharper than the inequality (1.15), it is the latter, as one would expect, that is more commonly used. (In Chapter 4 we shall investigate the sharpness of (1.14) itself.) It is also condition (1.15) that captures, more clearly, the notion that the continuity of the process is a consequence of the smoothness of the covariance function at the origin.

It is important to note, however, that in the move from an explicit entropy condition like (1.3) to an implicit one of the form (1.14), we may have lost something. For stationary Gaussian processes on $\Re^1$, this is not the case (as we shall see later). In general however, if X_t is continuous (even stationary) on $T \subset \Re^1$ and satisfies (1.14), then it can be transformed by a homeomorphism f of T into a process $X_{f(t)}$ which may no longer satisfy (1.14) but will, of course, still be continuous. Such a transformation may not preserve stationarity, and so does not contradict the claim of the previous sentence. Furthermore, (as you should check for yourself – Exercise 3.1) a transformation of this kind has no effect on the entropy of X, and so its continuity can still be checked via entropy methods.

With the continuity question more or less settled, a natural question to ask at this stage is how to construct discontinuous processes on $\Re^1$. To do this, let us impose the restriction that X be stationary. In this case the function $p(u)$ simplifies to give

$$p^2(u) = \sup_{0 \le t \le u} E|X_t - X_0|^2 = 2 \sup_{0 \le t \le u} [R(0) - R(t)], \tag{1.17}$$

and the finiteness of the integral in (1.14) becomes a necessary, as well as sufficient, condition for sample path continuity. (See Chapter 4 for details.)

It is a straightforward exercise, given the results of Chapter 4 (see Exercise 3.3), to show that if

$$\frac{C_1}{|\log t|^{1+\alpha_1}} \le R(0) - R(t) \le \frac{C_2}{|\log t|^{1+\alpha_2}}, \tag{1.18}$$

for $|t|$ small enough, then X will be sample path continuous if $\alpha_2 > 0$ and discontinuous if $\alpha_1 < 0$.

For the sake of the reader who is familiar with the spectral theory of stationary processes, the continuity/discontinuity dichotomy can be most naturally stated in spectral terms.

Thus, let $F(\lambda)$ denote the spectral distribution function defined by

$$R(t) = \int_0^\infty \cos t\lambda \, dF(\lambda), \tag{1.19}$$

with spectral density $f(\lambda) = F'(\lambda)$, when F is absolutely continuous.

Via a reasonably standard Tauberian theorem (e.g. Cramér and Leadbetter (1967) pp. 174-177) it follows from (1.18) that if the integrals

$$\int_0^\infty \left(\log(1+\lambda)\right)^{1+\alpha} dF(\lambda) = \int_0^\infty \left(\log(1+\lambda)\right)^{1+\alpha} f(\lambda)\, d\lambda \tag{1.20}$$

converge for some $\alpha > 0$ then X is continuous, while if they diverge for some $\alpha < 0$ then X has discontinuous sample paths with probability one.

Thus, if you are familiar with the spectral approach, it is now obvious from (1.20) exactly how much "high frequency oscillation" is allowed before a stationary Gaussian process on the line becomes discontinuous. If you are not familiar with spectra, there is no need to worry. They will not appear again.

Exactly what happens when a Gaussian process is discontinuous is a topic we shall return to in part below, in the example treating generalised processes, and in both of the following chapters.

2. GAUSSIAN FOURIER SERIES: A problem studied in the 1930's by Paley and Zygmund, and at first seemingly unrelated to Gaussian processes, is the question of the uniform convergence of the random Fourier series

$$\sum_{n=0}^{\infty} a_n Y_n e^{int}, \qquad t \in [0, 2\pi], \tag{1.21}$$

where the a_n are real numbers satisfying $\sum_{n=0}^{\infty} a_n^2 = 1$ and the Y_n are a sequence of independent Rademacher random variables; i.e. $P\{Y_n = +1\} = P\{Y_n = -1\} = \frac{1}{2}$. An equivalent (this requires some work to show) question arises when the Y_n of (1.21) are taken to be standard normal.

Both the Rademacher and Gaussian versions of this problem have a number of interesting consequences in non-random harmonic analysis, and it was for this reason that they were posed. Furthermore, the summands in (1.21) can be treated in much greater generality, with the parameter t ranging over a compact Abelian or non-Abelian group, and the non-random a_n replaced by a product of random and non-random terms. You can read about all of this in depth in the monograph of Marcus and Pisier (1981), but for now we shall treat only the simple sum (1.21), and assume that the Y_n are standard normal.

Apart from being of independent mathematical interest, (1.21) is of course interesting as a way to build a stationary Gaussian process on $[0, 2\pi]$. Looked at this way, it is not surprising that the results related to Gaussian processes on the line have something to say about Paley and Zygmund's problem.

Consider the following related version of the original problem, which is a little easier to handle in the context of the *real*-valued Gaussian processes that we are dealing with.

1.5 THEOREM. *Let $\{Y_n\}_{n\geq 1}$ and $\{Y_n'\}_{n\geq 1}$ be two independent, infinite sequences of independent, standard normal random variables, and $\{a_n\}_{n\geq 1}$ a non-increasing real sequence. Then the sum*

$$X_t := \sum_{n=0}^{\infty} a_n (Y_n \cos nt + Y_n' \sin nt), \qquad t \in [0, 2\pi], \tag{1.22}$$

converges uniformly on $[0, 2\pi]$ *if, and only if, the following sum also converges:*

$$(1.23) \qquad \sum_{j=2}^{\infty} \frac{\left(\sum_{n=j}^{\infty} a_n^2\right)^{1/2}}{j(\log j)^{1/2}} .$$

(An identical result, for the sum (1.21), can also be shown to be true. Such a result is actually more general than our formulation.)

PROOF: Actually, we shall only give a partial proof, as we shall only sketch why (1.23) is sufficient for uniform convergence. After you have read Chapter 3 – in particular Theorem 3.8 – you will be able to complete this part of the proof for yourself. (Exercise 2.5 of Chapter 3.) Despite the long wait, it is interesting to see, already at this stage, what is needed. The necessity is similar, but a little harder. You can find it in Marcus and Pisier (1981) or whichever of the references there to earlier, less abstract, results that you find most convenient.

Since we are assuming the convergence of (1.23), we have that the sequence $\{a_n\}_{n \geq 1}$ is square summable. Thus, by the three series theorem, the sum (1.22) converges for each *fixed* $t \in [0, 2\pi]$, with probability one. Let it do so on a countable dense subset of $[0, 2\pi]$, which, by separability, defines a limit process. The limit is clearly Gaussian, since the Y_n and Y_n' are both Gaussian. What we now have to show is that this convergence is uniform on the parameter space $[0, 2\pi]$. This would ordinarily require showing the existence of a common uniform modulus of continuity for the partial sums defining X_t. The results of Section 3.2, however, are rich in zero-one and related laws that tell us that all we need check is that the limit process, which we have just defined, is continuous. Uniform convergence then comes for free.

To prove the requisite continuity, we start by noting that an easy calculation shows the covariance function of the process X_t of the theorem to be

$$R(s,t) = R(t-s) = \sum_{n=0}^{\infty} a_n^2 \cos\left(n(t-s)\right).$$

The function p of (1.13) is thus given by

$$(1.24) \quad p^2(u) = 2 \sup_{0 \leq t \leq u} \sum_{n=0}^{\infty} a_n^2 (1 - \cos nt) = 4 \sup_{0 \leq t \leq u} \sum_{n=0}^{\infty} a_n^2 \sin^2\left(\frac{nt}{2}\right).$$

We now have to show that, with this formulation of p, the entropy integral (1.14) converges when the sum (1.23) does. Using Exercise 3.2, it is easy to see that the entropy integral converges if

$$(1.25) \qquad \sum_{n=0}^{\infty} \frac{p(2^{-n})}{\sqrt{n}} < \infty.$$

Since $\sin^2 x < \max(1, x^2)$ for all x, we have from (1.24) that

$$p^2(2^{-n}) \le 4\sum_{j=0}^{n} \frac{2^{2j}}{2^{2n}} A(2^j, 2^{j+1}) + B(2^n),$$

where

$$A(n,m) = \sum_{j=n+1}^{m} a_j^2, \qquad B(n) = A(n,\infty) = \sum_{j=n+1}^{\infty} a_j^2.$$

Using $(|x| + |y|)^{1/2} \le |x|^{1/2} + |y|^{1/2}$ we see that the sum in (1.25) will converge if both

$$\sum_{n=0}^{\infty} \Big(\frac{2^{-2n}}{n} \sum_{j=0}^{n} 2^{2j} A(2^j, 2^{j+1})\Big)^{\frac{1}{2}} < \infty, \tag{1.26}$$

and

$$\sum_{n=0}^{\infty} \Big(\frac{B(2^n)}{n}\Big)^{\frac{1}{2}} < \infty. \tag{1.27}$$

That (1.27) follows from the convergence of (1.23) is easy to check. It is also true that (1.26) follows from (1.27), but here the proof relies on an old inequality for series that takes a lot of time to prove and is of little independent interest as far as Gaussian processes are concerned. Hence we leave this out, and refer the reader to Marcus and Pisier (1981), once again, for full details. ∎

Although we have perhaps left more unproven than proven, we now leave this example and turn to some others to continue building motivation for the general theory to come.

3. THE TALAGRAND EXPANSION: Whereas we initially posed the question of the convergence of the random Fourier series of the previous example as a natural problem in harmonic analysis, from the point of view of a probabilist it is far more natural to view (1.22), for example, as an orthogonal expansion of a Gaussian process on a finite interval. From this viewpoint, a natural question to ask is whether or not orthogonal expansions exist in general (i.e. when the parameter space T has no special structure). When X_t is continuous over T the answer to this question is in the affirmative, as we shall see in §3.2. Now, however, we consider a somewhat different expansion, due to Talagrand (1987).

Let $\{Y_n\}_{n\ge 0}$ be a sequence of (possibly dependent) centered Gaussian variables, with variances $\sigma^2(Y_n)$ satisfying

$$\lim_{n\to\infty} (\log n)^{\frac{1}{2}} \sigma(Y_n) = 0. \tag{1.28}$$

Thus, like the i.i.d. Y_n and Y_n' of (1.22), our new variables become smaller, stochastically, as $n \to \infty$. (To be precise, in the previous examples the size of the random summands was governed indirectly through the non-random coefficients a_n.)

Let T be a compact metric space, and $\{\alpha_n\}_{n \geq 0}$ a summable sequence of positive functions on T such that

$$\alpha_n(t) \geq 0, \qquad \sum_{n=0}^{\infty} \alpha_n(t) \leq 1, \tag{1.29}$$

for all $t \in T$. Define the centered Gaussian process

$$X_t = \sum_{n=0}^{\infty} \alpha_n(t) Y_n. \tag{1.30}$$

It is not too hard to show (Exercise 3.4) that, under the conditions we have specified, X_t is sample path continuous. This is half of the following theorem. What is rather surprising, however, is the converse, which is also much harder to prove.

1.6 THEOREM. *Let X be a centered Gaussian process on a compact metric space T. Then X is continuous if, and only if, it has a continuous covariance function, there exists a Gaussian sequence $\{Y_n\}_{n \geq 0}$ satisfying (1.28), and, for each $t \in T$, X_t can be written in the form (1.30), where the series converges in $\mathcal{L}^2(P)$ and the α_n satisfy (1.29).*

The practical power of this result is that it gives us a very simple, and very general, method for constructing continuous Gaussian processes, particularly in situations where it might be difficult to write down an orthonormal basis for functions on T. Furthermore, it tells us that *all* continuous processes can be built this way.

Two aspects of the theorem are of particular interest. Firstly, it is important to note that the Y_n need not be independent. This somewhat limits the usefulness of the converse part of the result. Secondly, nowhere have we made the (unnecessary) requirement that the α_n be continuous. For more details, including how to prove the easy half of Theorem 1.6, see Exercise 3.4.

4. GENERALISED RANDOM FIELDS:

The term "random field" is generally used to describe a stochastic process whose parameter space is either a k-dimensional Euclidean space or a k-dimensional lattice. Thus, a 1-dimensional random field is what is generally referred to as a "stochastic process". Problems related to the geometrical structure of the sample paths of random fields ($k \geq 1$) are often qualitatively different, and substantially more difficult, than the corresponding problems when $k=1$: c.f. Adler (1981). Problems related to continuity, however, are generally dimension independent. For example, Theorem 1.4 remains true if we merely replace the domain of X by a compact subset of $\Re^k$, and the definition (1.13) of the function $p(u)$ by

$$p^2(u) = \sup_{\|s-t\| \le \sqrt{k}u} E|X_s - X_t|^2, \tag{1.31}$$

where $\| \; \|$ is the usual Euclidean norm (Exercise 3.5). Similar changes can be made to (1.15) and (1.20): i.e. simply make t or λ vectors and change moduli signs to Euclidean norms where appropriate. Formula (1.17) changes by taking the supremum over $\{t\colon \|t\| \le \sqrt{k}u\}$.

Our interest in this section, therefore, will be not with random fields of the above kind, but rather with a class of Gaussian processes defined on certain function spaces. Nevertheless, we shall use simple random fields to provide an example of the processes that do interest us.

Thus, let X_t be a centered, Gaussian random field on $\Re^k$, with covariance function $R(s,t)$. Let $\mathcal{F}$ be a family of functions on $\Re^k$, and for $\phi \in \mathcal{F}$ define

$$X(\phi) = \int_{\Re^k} \phi(t)\, X(t)\, dt. \tag{1.32}$$

We thus obtain a centered Gaussian process indexed by functions in $\mathcal{F}$, whose covariance function(al) is given by

$$R(\phi,\psi) = EX(\phi)\, X(\psi) = \int_{\Re^k}\int_{\Re^k} \phi(s) R(s,t) \psi(t)\, dsdt. \tag{1.33}$$

The main interest in (1.33) is that it often allows us to define function indexed Gaussian processes with covariance functional of this form, even when a point indexed process X_t with covariance function $R(s,t)$ does not exist. The examples that we shall consider here arise when $R(t,t) \equiv \infty$ for all t, and the divergence near the diagonal is bounded as follows:

$$R(s,t) \le \frac{C}{\|s-t\|^\alpha}, \qquad \text{for all} \quad \|s-t\| \le \delta, \tag{1.34}$$

for some $C < \infty$ and $\delta > 0$.

The main questions of interest here are on what families $\mathcal{F}$ can a process satisfying (1.33) be defined, and when is it continuous.

The answer to the first question is trivial: Given any positive definite function R, one can define a function indexed process on

$$\mathcal{F}_R = \{\phi\colon \int_{\Re^k}\int_{\Re^k} \phi(s) R(s,t) \phi(t)\, dsdt < \infty\}. \tag{1.35}$$

The answer to the second question involves a considerable amount of work, but you should follow the argument carefully, because it is one of the few places in these notes that we shall give an involved entropy calculation in a reasonably complicated example in reasonably complete detail.

As to why we have chosen covariances satisfying (1.34), and how these are related to wider classes of function (and measure) indexed functions, see both the comments at the end of this collection of examples, and the examples in the following subsection.

Let $T \subset \Re^k$ be bounded, $q > 0$, and $p = [q]$. Let $C_0, \ldots, C_p$ and C_q be finite, positive constants, and let $\mathcal{F}^{(q)} = \mathcal{F}^{(q)}(T, C_0, \ldots, C_p, C_q)$ be the class of functions on T whose partial derivatives of orders $1, \ldots, p$ are bounded by $C_0, \ldots, C_p$, and for which the partial derivatives of order p satisfy Hölder conditions of order $q - p$ with constant C_q. Thus for each $\phi \in \mathcal{F}^{(q)}$ and $t, t+\tau \in T$

$$\phi(t+\tau) = \sum_{n=0}^{p} \frac{\Phi_n(t,\tau)}{n!} + \Delta(t,\tau), \tag{1.36}$$

where $\Phi_n(t,\tau)$ is a homogeneous polynomial of degree n in τ of the form

$$\Phi_n(t,\tau) = \Phi_n(t: \tau_1, \ldots, \tau_k) = \sum_{j_1=1}^{k} \cdots \sum_{j_n=1}^{k} \frac{\partial^n \phi(t)}{\partial t_{j_1} \ldots \partial t_{j_n}} \tau_{j_1} \cdots \tau_{j_n}, \tag{1.37}$$

and where

$$\sup_{t\in T} \left| \frac{\partial^n \phi(t)}{\partial t_{j_1} \ldots \partial t_{j_n}} \right| \le C_n \quad \text{and} \quad |\Delta(t,\tau)| \le C_q \|\tau\|^q. \tag{1.38}$$

Two things are obvious in the above setup, in which you should think of the dimension k as fixed. Firstly, the larger the α in (1.34) the rougher the process with this covariance will be. Secondly, the larger q is the smaller the family $\mathcal{F}^{(q)}$ will be, and thus the more likely that a Gaussian process defined on $\mathcal{F}^{(q)}$ will be continuous. Thus it seems reasonable to expect that a result of the following kind should be true. (For a partial result in the converse direction, see Example 3.6.)

1.7 THEOREM. *A centered Gaussian process with covariance function satisfying (1.33) and (1.34) will be continuous on* $\mathcal{F}^{(q)}(T, C_0, \ldots, C_p, C_q)$ *if* $k > \alpha$ *and if the following condition is satisfied:*

$$q > \frac{1+\alpha-k}{2}.$$

REMARKS: Note that since we have not specified any other metric on $\mathcal{F}^{(q)}$, the continuity claim of the Theorem is with relation to the topology induced by the canonical metric. There are, of course, more natural metrics on $\mathcal{F}^{(q)}$, but as long as we are interested in mere continuity and not in more detailed information such as moduli of continuity we need not bother with them.

If, despite the advice given above, you want to let k vary in the condition of the Theorem, while keeping α and q fixed, you will find that the larger k is, the less derivatives we require of our test functions to ensure continuity on $\mathcal{F}^{(q)}$. While at first this seems counter-intuitive, you should remember that as k increases the degree of the singularity in (1.34) decreases (for fixed α) and so the result is, in fact, reasonable.

While the reason for the assumption $k > \alpha$ should be obvious from the proof, it is worthwhile noting already that it is precisely this condition that gives us a process with finite variance, since, for $\phi \in \mathcal{F}^{(q)}$ with $\|\phi\|_\infty < M$

$$\begin{aligned} EX^2(\phi) &= \int_T \int_T \phi(s)R(s,t)\phi(t)\, dsdt \\ &\le M^2 \int_T \int_T R(s,t)\, dsdt \\ &\le CM^2 \int_T \int_T \|t-s\|^{-\alpha}\, dsdt. \end{aligned}$$

Since $T \subset \Re^k$ is compact, a transformation to polar coordinates easily shows that the last integral is finite only if $k > \alpha$.

PROOF: To make life notationally easier, we shall assume throughout that $T = [0,1]^k$.

The proof is, unfortunately, rather long. It will proceed via an entropy argument. Thus what we need to show, in the light of Theorem 1.1, is that the entropy of $\mathcal{F}^{(q)}$, as measured in the canonical metric d, where

$$d^2(\phi,\psi) = \int_{\Re^k} \int_{\Re^k} (\phi(s) - \psi(s))\, R(s,t)\, (\phi(t) - \psi(t))\, dsdt,$$

is small enough for the entropy integral (1.3) to converge. We shall do this by explicitly constructing, for each $\epsilon > 0$, a finite family $\mathcal{F}_\epsilon^{(q)}$ of functions that serve as an ϵ-net for $\mathcal{F}^{(q)}$ in the d metric.

We start by fixing $\epsilon > 0$ and defining

$$\delta = \delta(\epsilon) = \epsilon^{1/(q+(k-\alpha)/2)}. \tag{1.39}$$

Let Z_δ denote the grid of the $(1 + [\delta^{-1}])^k$ points in $[0,1]^k$ of the form

$$t_\eta = (\eta_1\delta, \ldots, \eta_k\delta), \qquad \eta_i = 0, 1, \ldots, [\delta^{-1}], \quad i = 1, \ldots, k. \tag{1.40}$$

Define

$$\delta_n = \delta_n(\epsilon) = \delta^{q-n}, \qquad n = 0, \ldots, p,$$

and for each $\phi \in \mathcal{F}^{(q)}$, $n = 0, \ldots, p$, and t_η of the form (1.40) let $A_\eta^{(n)}(\phi)$ denote the vector formed by taking the integer part of δ_n^{-1} times the partial derivatives of ϕ of order n evaluated at the point t_η. (The index η here is, of course, k-dimensional.) Thus, a typical element of $A_\eta^{(n)}$ is of the form

$$(1.41) \qquad \left(A_\eta^{(n)}(\phi)\right)_i = \left[\frac{\phi^{(n_1,\ldots,n_k)}(t_\eta)}{\delta_n}\right], \qquad n_1 + \cdots + n_k = n,$$

where we have written $\phi^{(n_1,\ldots,n_k)}$ for the derivative $\partial^n \phi / \partial^{n_1} \ldots \partial^{n_k}$, and the index i runs from 1 to $\binom{n+k-1}{k-1}$.

Finally, for $\phi \in \mathcal{F}^{(q)}$, let $A_\delta = A_\delta(\phi)$ denote the $(k+1)$-dimensional matrix, a generic element of which is the vector $A_\eta^{(n)}$, $t_\eta \in Z_\delta$, $n = 0, \ldots, p$, (η is k-dimensional, n is one-dimensional) and let $F_{A(\delta)}$ denote the set of $\phi \in \mathcal{F}^{(q)}$ with fixed matrix A_δ. Our first task will be to show that the d-radius of $F_{A(\delta)}$ is not greater than $C\epsilon$, where C is a constant dependent only on q and k. All that will then remain will be to calculate how many different collections $F_{A(\delta)}$ are required to cover $\mathcal{F}^{(q)}$.

Thus, take $\phi_1, \phi_2 \in A_\delta$, and set

$$(1.42) \qquad \phi = \phi_1 - \phi_2.$$

Let $\|\ \|_d$ be the norm induced on $\mathcal{F}^{(q)}$ by the metric d, and $\|\ \|_\infty$ the usual sup norm. Thus,

$$\|\phi\|_d^2 = \int_{[0,1]^k} \int_{[0,1]^k} \phi(s) R(s,t) \phi(t)\, ds dt, \qquad \|\phi\|_\infty = \sup_{[0,1]^k} |\phi(t)|.$$

We have to show that the ϕ of (1.42) has d-norm less than $C\epsilon$.

Note first, however, that in view of the definition of the matrix A_δ via (1.41), we have that for each $t_\eta \in Z_\delta$ and each partial derivative $\phi^{(n_1,\ldots,n_k)}$ of such a ϕ of order $n_1 + \cdots + n_k = n \le p$ that

$$\phi^{(n_1,\ldots,n_k)}(t_\eta) \le \delta_n.$$

(Don't forget that the square brackets in (1.41) mean "integer part".)

Putting this inequality together with the Taylor expansion (1.36) and (1.37), we find that for all $t \in [0,1]^k$

$$\begin{aligned} |\phi(t)| &\le \sum_{n=0}^{p} \frac{k^n}{n!} \delta_n \delta^n + C\delta^q \\ &= C(k,p)\delta^q, \end{aligned}$$

the last line following from the definition of the δ_n and the fact that each polynomial Φ_n of (1.37) has less than k^n distinct terms.

Thus, for ϕ of the form (1.42),

$$\|\phi\|_\infty \leq C\delta^q. \tag{1.43}$$

We now turn to $\|\phi\|_d$. With δ as above, set

$$D_\delta = \{(s,t) \subset [0,1]^k \times [0,1]^k : \max_{i=1,\dots,k} |s_i - t_i| \leq \delta\}.$$

Then

$$\begin{aligned} \|\phi\|_d^2 &= \int_{[0,1]^k \times [0,1]^k} \phi(s)R(s,t)\phi(t)\, dsdt \\ &= \int_{D_\delta} \phi(s)R(s,t)\phi(t)\, dsdt \\ &\qquad + \int_{([0,1]^k \times [0,1]^k)\setminus D_\delta} \phi(s)R(s,t)\phi(t)\, dsdt \\ &= I_1(\delta) + I_2(\delta). \end{aligned} \tag{1.44}$$

Consider the first integral. Letting C change from line to line where necessary, we have from (1.34) and (1.43) that

$$\begin{aligned} I_1(\delta) &\leq C\,\delta^{2q} \int_{[0,1]^k} ds \int \cdots \int_{s_i-\delta \leq t_i \leq s_i+\delta} dt\, \|s-t\|^{-\alpha} \\ &\leq C\,\delta^{2q} \int_{[-\delta,\delta]^k} \|t\|^{-\alpha}\, dt \\ &\leq C\,\delta^{(2q+k-\alpha)}, \end{aligned} \tag{1.45}$$

the last inequality coming from an evaluation via polar coordinates, and requiring the condition $k > \alpha$.

Similarly, again relying on the fact that $k > \alpha$, it is easy to check that $I_2(\delta)$ is also bounded above by $C\,\delta^{(2q+k-\alpha)}$. Substituting this fact and (1.45) into (1.44), and applying (1.39), we finally obtain that for ϕ satisfying (1.42)

$$\|\phi\|_d < C\,\delta^{\frac{1}{2}(2q+k-\alpha)} = C\epsilon. \tag{1.46}$$

That is, the d-radius of each set $A_\eta^{(n)}$ is no greater than a uniform constant times ϵ.

It remains to determine how many collections $F_{A(\delta)}$ are required to cover $\mathcal{F}^{(q)}$. Since this is a calculation that is now independent of both Gaussian processes in general, and the above covariance function in particular, we shall only outline how this is done. The details, which require somewhat cumbersome notation, can be found in Kolmogorov and Tihomirov (1959).

Consider, for fixed δ, the $(k+1)$-dimensional matrix A_δ, parameterised, as in (1.40), by $\eta_i = 0, 1, \ldots, [\delta^{-1}]$, $i = 1, \ldots, k$, and $n = 0, 1, \ldots, p$. Fix, for the moment, $\eta_2 = \cdots = \eta_k = 0$. It is clear from the restrictions (1.38), (1.41), the definition of δ_n, and the fact that each vector $A_\eta^{(n)}$ has no more than $\binom{n+k-1}{k-1}$ distinct elements, that there are no more than

$$O\left(\frac{1}{\delta_0}\left(\frac{1}{\delta_1}\right)^{\binom{k}{k-1}} \cdots \left(\frac{1}{\delta_p}\right)^{\binom{p+k}{k-1}}\right) = O(\delta^{-\xi}),$$

(for an appropriate and eventually unimportant ξ) ways to fill in the row of A_δ corresponding to $(n_1, \ldots, n_k) = (0, \ldots, 0)$.

The main part of the proof is to now show that because of the rigid continuity conditions on the functions in $\mathcal{F}^{(q)}$, there exists an absolute constant $M = M(q, C_0, \ldots, C_p, C_q)$, such that once this first row is determined, there are no more than M ways to complete the row corresponding to $(n_1, \ldots, n_k) = (1, \ldots, 0)$, and similarly no more than M^2 ways to complete the row corresponding to $(n_1, \ldots, n_k) = (2, \ldots, 0)$, etc. Thus, all told, there are no more than

$$O\left(\delta^{-\xi} \cdot M^{k(1+\delta^{-1})}\right) \tag{1.47}$$

ways to fill the matrix A_δ, and thus we have a bound for the number of different collections $F_{A(\delta)}$.

Modulo a constant, it now follows from (1.39), (1.46) and (1.47) that the logarithm of the entropy function for our process is bounded above by

$$C_1 + \frac{C_2\,\xi}{(q + (k-\alpha)/2)} \log\left(\frac{1}{\epsilon}\right) + C_3\left(\frac{1}{\epsilon}\right)^{1/(q+(k-\alpha)/2)}. \tag{1.48}$$

Substituting this into the entropy condition (1.3) (noting that it is really only the last term above that is important) is all that is now required to complete the proof of the theorem. ∎

Before leaving this example, there are a number of comments that are worth making, that relate it to other problems both within and outside of the theory of Gaussian processes.

Firstly, in most of the literature pertaining to generalised Gaussian fields the parameter space used is the Schwartz space $\mathcal{S}$ of infinitely differentiable functions decaying faster than any polynomial at infinity. Since this is a very small class of functions (at least in comparison to the classes $\mathcal{F}^{(q)}$ that Theorem 1.7 deals with) continuity over $\mathcal{S}$ is automatically assured and therefore not often explicitly treated. However, considerations of continuity and smaller parameter spaces are of relevence in the treatment of infinite dimensional diffusions, and associated infinite dimensional stochastic partial differential equations, in which solutions over very specific parameter spaces are often sought. For more on this see, for example, Itô (1983) and Watanabe (1984).

Secondly, some words on our choice of (1.34) as a condition on the covariance kernel $R(s,t)$. When $\alpha = k - 2$, $k > 2$ then the class of generalised fields that we are considering here includes the so called "free field" of Euclidean quantum field theory. (When $k = 2$ the free field has a covariance kernel with a logarithmic singularity at 0, and when $k = 1$ the free field is no longer generalised, but is the stationary Markov Gaussian process with covariance function $R(t) = e^{-\beta|t|}$, for some $\beta > 0$.) This process, along with a large number of related generalised fields whose covariance kernels satisfy similar conditions, possesses a type of multi-dimensional Markov property. For details on this see, for example, Dynkin (1980, 1984), Adler and Epstein (1987), and references therein. For structural and renormalisation properties of generalised fields of this kind, presented among a much wider class of examples, see Dobrushin (1979), who also treats a large variety of non-Gaussian fields. We shall have a little more to say about some of these processes in Chapter 6, where we shall discuss how they are related to $\Re^d$-valued Markov processes on the real line.

Finally, it is worth noting that much of what has been said above regarding generalised fields – i.e. function indexed processes – can be easily extended to Gaussian processes indexed by a family of measures. For example, if we consider the function ϕ in (1.32) to be the (positive) density of a measure μ on $\Re^k$, then by analogy with (1.32) it makes sense to write

$$X(\mu) = \int_{\Re^k} X(t)\,\mu(dt),$$

with the corresponding covariance functional

$$R(\mu,\nu) = EX(\mu)X(\nu) = \int_{\Re^k}\int_{\Re^k} \mu(ds)R(s,t)\nu(dt).$$

Again, as was the case for generalised Gaussian fields, the process $X(\mu)$ may be well defined even if the covariance kernel R diverges on the diagonal. In fact, $X(\mu)$ will be well defined for all $\mu \in \mathcal{M}_R$, where

$$\mathcal{M}_R = \{\mu\colon \int_{\Re^k}\int_{\Re^k} \mu(ds)R(s,t)\mu(dt) < \infty\}.$$

Similar arguments to those used above to characterise the continuity of a family of Gaussian fields on $\mathcal{F}^{(q)}$ can be used to ascertain continuity of measure indexed processes on suitably smooth classes of measures. We leave both the details and an attempt to formulate the appropriate results to the interested reader. (Actually, to the best of my knowledge, this specific example has never been treated in the literature, and so it would be a rather interesting problem to work out how to optimally, and naturally, formulate the requisite smoothness conditions. The general idea of how to proceed can be gleaned from the treatment of the following example.)

5. SET INDEXED PROCESSES:

We have already met some set indexed processes in dealing with the Brownian family of processes in the previous section. The motivation behind studying those processes lay in the development of multivariate Kolmogorov-Smirnov tests based on statistics of the type (1.11). Our aim now will be merely to introduce two new families $\mathcal{A}$ of subsets of $\Re^k$ that have proven useful as parameter spaces for the Gaussian white noise W satisfying (1.5)–(1.7), so that

$$EW(A)W(B) = R_\nu(A,B) = \nu(A \cap B), \qquad A,B \in \mathcal{A}, \tag{1.49}$$

where ν is a σ-finite measure on $\Re^k$.

Before we start in earnest, however, it is worthwhile taking a moment to consider the particularly simple family of special cases obtained when we take $\mathcal{A}$ to be a family of n-sided polygons in $\Re^2$. (A similar argument will apply to n-sided polyhedra in higher dimensions.)

To simplify the notation, let $\mathcal{A}$ be the family of all rectangular subsets of $[0,1]^2$. A typical element of $\mathcal{A}$ is $A = [s,t] \times [u,v]$. The geometry of W on this $\mathcal{A}$ is non-trivial, and can be conceptually simplified by defining a new, point indexed, random field X on the set

$$\mathcal{D} = \{(s,t,u,v) \in [0,1]^4 : s \le t,\ u \le v\}$$

by the correspondence

$$X(s,t,u,v) = W\big([s,t] \times [u,v]\big).$$

Thus, the questions of sample path continuity and boundedness of W on $\mathcal{A}$ reduce to the same questions for the point indexed random field X on $\mathcal{D}$, and we have already seen, at the beginning of the subsection on generalised random fields, how to treat such cases.

However, as simple as this approach may sound, it is not always quite so straightforward to implement. The main problem is that whereas the covariance function of W on $\mathcal{A} \times \mathcal{A}$ is natural and easy to work with, the covariance of X on $\mathcal{D} \times \mathcal{D}$ has very little in the way of "geometric symmetry", and specific calculations can often be difficult, depending on the structure of ν. A secondary problem is that some of the clever arguments that we shall encounter in Chapter 5 for similarly indexed Brownian sheets – arguments that rely heavily on geometrical properties of $\mathcal{A}$ – break down on $\mathcal{D}$. Finally, of course, there is the problem that similar methods will not apply when a parameter space of sets does not have as simple a finite dimensional representation as do spaces of polyhedra.

It is to two such examples that we now turn.

The first family of sets is closely related to the family $\mathcal{F}^{(q)}$ of functions we have just studied in detail, and is, essentially, composed of those sets in $\Re^k$ that have q times differentiable boundaries. It is developed as follows:

Let S^{k-1} be the unit sphere in $\Re^k$, so that

$$S^{k-1} = \{t \in \Re^k : \|t\|^2 = 1\}.$$

Recall the basic result from differential topology that we can cover S^{k-1} by finitely many coordinate patches V_j, each represented as a C^∞ isomorphism $F_j : B^{k-1} \to V_j$ of the open ball $B^{k-1} = \{t \in \Re^{k-1} : \|t\|^2 < 1\}$.

Adapting slightly the notation of the previous example, let $\mathcal{F}^{(q)}(V_j, M)$ be the set of all real valued functions ϕ on V_j such that $\phi \circ F_j \in \mathcal{F}^{(q)}$ $(B^{k-1}, M, \dots, M)$ (c.f. (1.36)–(1.38)). Furthermore, let $\mathcal{F}^{(q)}(S^{k-1}, M)$ denote the set of all real valued functions ϕ on S^{k-1} such that the restriction of ϕ to V_j is in $\mathcal{F}^{(q)}(V_j, M)$. Taking the k-fold Cartesian product of copies of $\mathcal{F}^{(q)}(S^{k-1}, M)$, we obtain a family of functions from S^{k-1} to $\Re^k$, which we denote by $D(k,q,M)$, where the "D" stands for "Dudley", who introduced this family in Dudley (1974).

Each $\phi \in \mathcal{F}^{(q)}$ defines a $(k-1)$-dimensional surface in $\Re^k$, and a simple algebraic geometric construction enables one to "fill in" the interior of this surface to obtain a set I_ϕ. We shall denote the family of sets obtained in this fashion by $I(k,q,M)$, and call them the "Dudley sets with q-times differentiable boundaries".

Now we specialise to the special case in which the measure ν in (1.49) is Lebesgue measure, so that the process W there becomes the Brownian sheet. Calculations, whose nature we shall discuss in the proof outline below, then show that the log-entropy function for the Brownian sheet on $I(k,q,M) \cap T$, where $T \subset \Re^k$ is compact, satisfies

(1.50)

$$\log N\big(I(k,q,M),\epsilon\big) \le \begin{cases} C_T \left(\dfrac{1}{\epsilon}\right)^{2(k-1)/(kq-k+1)} & (k-1)/k < q \le 1, \\ C_T \left(\dfrac{1}{\epsilon}\right)^{2(k-1)/q} & 1 \le q. \end{cases}$$

(Equivalent lower bounds for the log-entropy are also available for certain values of k and q.)

As an immediate conseqence of the above and Theorem 1.1 we have

1.8 PROPOSITION. *The Brownian sheet is continuous on a bounded collection of Dudley sets in* $\Re^k$ *with* q *times differentiable boundaries if* $q > k-1 \geq 1$. *If* $k-1 > 1 \geq q > 0$ *or if* $k-1 > q \geq 1$ *then the Brownian sheet is unbounded with probability one.*

OUTLINE OF PROOF: The proof of the unboundedness part of the result is beyond us at the moment. As far as the proof of continuity is concerned, all that really needs to be proven are the entropy inequalities above. These rely the "simple algebraic geometric construction" noted above, and so we shall not bring them in detail. The basic idea, however, requires little more than noting that there are basically as many sets in $I(k,q,M)$ as there are are functions in $D(k,q,M)$, and we have already seen, in the previous example, how to count the number of functions in $D(k,q,M)$. You can find the details in Dudley (1974), or try it for yourself in Example 3.7. ■

The astute reader will have noticed that whereas both in §1.2 and above we defined Gaussian white noise via any σ-finite measure ν (c.f. (1.49)) up until now we have only discussed the Brownian sheet based on Lebesgue measure. If we wish to extend the discussion to more general measures, and retain properties of continuity, we need to substantially restrict the class of sets $\mathcal{A}$ under consideration as the parameter space.

The most common classes of sets that arise in this context are the so called Vapnick-Červonenkis, or VC, sets, due, not surprisingly, to Vapnick and Červonenkis (1971). Since the arguments involved in entropy calculations are of an essentially combinatoric nature, and so of a somewhat different form to those we have met before, we shall now discuss them in some detail. (For a much more substantial treatment, including the importance of VC classes to the problem of finding "universal Donsker classes" in the theory of empirical processes, see, for example, Dudley (1978a).)

Let E be a subset of $\Re^k$, and ν a probability measure on E. Given a class $\mathcal{C}$ of subsets of E and a finite set $F \subset E$, let $\Delta^{\mathcal{C}}(F)$ be the number of different sets $C \cap F$ for $C \in \mathcal{C}$. For $n = 1, 2, \ldots$, let

$$m^{\mathcal{C}}(n) := \max\{\Delta^{\mathcal{C}}(F) \colon F \text{ has } n \text{ elements}\}.$$

Clearly, $m^{\mathcal{C}}(n) \leq 2^n$ for all n. Also, set

$$(1.51) \qquad V(\mathcal{C}) = \begin{cases} \inf\{n \colon m^{\mathcal{C}}(n) < 2^n\}, & \text{if } m^{\mathcal{C}}(n) < 2^n \text{ for some } n. \\ \infty, & \text{if } m^{\mathcal{C}}(n) = 2^n \text{ for all } n. \end{cases}$$

The class $\mathcal{C}$ is called a Vapnik-Červonenkis class if $m^{\mathcal{C}}(n) < 2^n$ for *some* n; i.e. if $V(\mathcal{C}) < \infty$. The number $V(\mathcal{C})$ is called the VC index of $\mathcal{C}$.

Two extreme but easy examples which you can check for yourself are $E = \Re$ and $\mathcal{C}$ all half lines, for which $m^{\mathcal{C}}(n) = n+1$ and $V(\mathcal{C}) = 1$, and $E = [0,1]$ with $\mathcal{C}$ all the open sets in $[0,1]$. Here $m^{\mathcal{C}}(n) = 2^n$ for all n and so $V(\mathcal{C}) = \infty$ and $\mathcal{C}$ is not a VC class.

A more instructive example, that also leads into the general theory we are after, is $E = \Re^k$ and $\mathcal{C}$ is the collection of half-spaces of $\Re^k$. Let $\Phi(k,n)$ be the maximal number of components into which it is possible to partition $\Re^k$ via n hyperplanes. Then, by definition, $m^{\mathcal{C}}(n) = \Phi(k,n)$. It is not hard to see that Φ must satisfy the following recurrence relation:

$$(1.52) \qquad \Phi(k,n) = \Phi(k,n-1) + \Phi(k-1,n-1),$$

with the boundary conditions $\Phi(0,n) = \Phi(k,0) = 1$. (If $\Re^k$ has already been partitioned into $\Phi(k,n-1)$ subsets via $n-1$ (($k-1$)-dimensional) hyperplanes, $H_1, \dots, H_{n-1}$, then adding one more hyperplane H_n will cut in half as many of these subsets as intersect H_n. There can be no more such subsets, however, than the maximal number of subsets formed on H_n by partitioning with the $n-1$ ($k-2$)-dimensional hyperplanes $H_1 \cap H_n, \dots, H_{n-1} \cap H_n$; i.e. $\Phi(k-1, n-1)$. Hence (1.52).)

Induction then shows that

$$(1.53) \qquad \Phi(k,n) = \begin{cases} \sum_{j=0}^{k} \binom{n}{j} & \text{if } n > k, \\ 2^n & \text{if } n \leq k, \end{cases}$$

where we adopt the usual convention that $\binom{n}{j} = 0$ if $n<j$.

From either the above or (1.52) you can now check that

$$(1.54) \qquad \Phi(k,n) \leq n^k + 1, \qquad \text{for all } k, n > 0.$$

It thus follows, from (1.51), that the half-spaces of $\Re^k$ form a VC class for all k. ■

What is somewhat more surprising, however, is that an inequality akin to (1.54), which we developed only for this special example, holds in general.

1.9 LEMMA. *Let E be any set, $\mathcal{C}$ any collection of subsets of E, and $V(\mathcal{C}) \leq v$. Then*

$$(1.55) \qquad m^{\mathcal{C}}(n) < \Phi(v,n) \leq n^v + 1, \qquad \text{for all } n \geq v.$$

Since the proof of this result is combinatoric rather than probabilistic, and will be of no further interest to us, you are referred to either Vapnick and Červonenkis (1971) or Dudley (1978a) for a proof.

The importance of Lemma 1.9 is that it enables us to obtain bounds on the entropy function for the Gaussian white noise based on *any* probability measure ν, and that these bounds are independent of ν.

1.10 THEOREM. *Let W be the Gaussian white noise based on a probability measure ν on some measure space $(E, \mathcal{E}, \nu)$. Let $\mathcal{A}$ be a Vapnik-Červonenkis class of sets in $\mathcal{E}$ with $V(\mathcal{A}) = v$. Then there exists a constant $K = K(v)$ (not depending on ν) such that for $0 < \epsilon \leq \frac{1}{2}$, the entropy function for W satisfies*

$$N(\mathcal{A}, \epsilon) \leq K\epsilon^{-2v} |\log \epsilon|^v.$$

PROOF: We start with a little counting, and then turn to the entropy calculation proper. The counting argument is designed to tell us something about the maximum number of $\mathcal{A}$ sets, that are a certain minimum distance from one another, that can be packed into E.

Suppose $A_1, \ldots, A_m \in \mathcal{A}$, $m \geq 2$, and $\nu(A_i \Delta A_j) \geq \epsilon$ for $i \neq j$. We need an upper bound on m. Sampling with replacement, select n points at random from E. The ν-probability that at least one of the sets $A_i \Delta A_j$ contains none of these n points is at most

$$\binom{m}{2}(1-\epsilon)^n. \tag{1.56}$$

Choose $n = n(m, \epsilon)$ large enough so that this probability is less than 1. Then

$$P\{\text{all symmetric differences } A_i \Delta A_j \text{ are non-empty}\} > 0,$$

and so for at least one configuration of the n sample points the class $\mathcal{A}$ picks out at least m distinct subsets. (Since, with positive probability, given any two of the A_i there is at least one point not in both of them.) Thus, by (1.55),

$$m \leq m^{\mathcal{A}}(n) \leq n^v = \big(n(m, \epsilon)\big)^v. \tag{1.57}$$

Take now the smallest n for which (1.56) is less than 1. For this n we have $m^2(1-\epsilon)^{n-1} \geq 2$, so that

$$n - 1 \leq \frac{2 \log m - \log 2}{|\log(1-\epsilon)|},$$

and $n \leq (2 \log m)/\epsilon$. Furthermore, by (1.57), $m \leq (2 \log m)^v \epsilon^{-v}$.

For some $m_0 = m_0(v) < \infty$, $(2 \log m)^v \leq m^{1/(v+1)}$ for $m \geq m_0$, and then $m \leq \epsilon^{-v-1}$, so $\log m \leq (v+1)|\log \epsilon|$. Hence

$$m \leq K(v)\, \epsilon^{-v} |\log \epsilon|^v \qquad \text{for } 0 < \epsilon \leq \tfrac{1}{2} \tag{1.58}$$

if $K(v) = \max(m_0, 2^{v+1}(v+1)^v)$.

This concludes the counting part of the proof. We can now do the entropy calculation. Recall that the canonical distance between on sets of $\mathcal{E}$ is given by $d_\nu(A,B) = [\nu(A\Delta B)]^{\frac{1}{2}}$.

Fix $\epsilon > 0$, and suppose that $A_1,\ldots,A_m$ are sets in $\mathcal{A}$ for which the union of ϵ-neighbourhoods of the A_i (in the d_ν metric) covers $\mathcal{A}$. By judicious trimming (see the construction of (4.15)–(4.16) for an example of the details) of the intersections of these neighbourhoods, one can assume that $d_\nu(A_i,A_j) \geq \epsilon$ for all i,j. Thus, by (1.58), $m \leq K(v)\epsilon^{2v}|2\log\epsilon|^v$, which proves the theorem. ■

An immediate consequence of the entropy bound of Theorem 1.10 is

1.11 COROLLARY. *Let W be the Gaussian white noise based on a probability measure ν on some measure space $(E,\mathcal{E},\nu)$. Then W is continuous over any Vapnik-Červonenkis class of sets in $\mathcal{E}$.*

With this we complete our two new examples of families of index sets for Gaussian processes.

You can find a lot more material on Vapnik-Červonenkis classes in the lecture notes by Dudley (1984), and some material for exercising your combinatorial skills in Examples 3.8–3.10. (If you find your combinatorial skills are not as sharp as you might like, look up Dudley's notes to find out how to solve the exercises.)

6. VECTOR VALUED PROCESSES:

Whereas the general theory espoused in the first two sections, and the examples we have looked at up until now, cover Gaussian processes on very general parameter spaces, all of our effort has been concentrated on real valued processes. It seems worthwhile to say at least a little about how well these results extend to Gaussian processes taking values in more general state spaces.

Thus, let $\mathbf{X}_t = \langle X_t^1,\ldots,X_t^N\rangle$ be a $\Re^N$-valued Gaussian process on a metric space (T,τ); i.e. each X_t^i is a real valued Gaussian process on (T,τ). The covariance function that has formed the basis of all our analysis up until now must be replaced with a matrix valued function $\mathbf{R}(s,t)$ with generic element $R_{ij}(s,t) = EX_s^iX_t^j$. We shall assume, as usual, that the X_t^i have zero mean.

Despite this complication, the question of the continuity of $\mathbf{X}$ is no harder than before. Since $\mathbf{X}$ is continuous as a function from (T,τ) to $(\Re^N,\|\ \|)$ if, and only if, each X_t^i is continuous as a real valued function on (T,τ), the global continuity question can be treated as N separate, real valued, problems.

The problem of the behaviour of the supremum of $\mathbf{X}$ is not quite as simple. If we concentrate our attention on the random variable

$$\sup_{t\in T}\|\mathbf{X}_t\| = \sup_{t\in T}\left((X_t^1)^2+\cdots+(X_t^N)^2\right)^{\frac{1}{2}}, \tag{1.59}$$

then it is clear that we are treating here the supremum of the non-Gaussian process $\|\mathbf{X}\|$. (If the N components of $\mathbf{X}$ are i.i.d. then $\|\mathbf{X}_t\|$ has a χ^2_N distribution. If they are independent but not identically distributed, then the distribution of $\|\mathbf{X}_t\|$ is non-central χ^2. In the general case, the distribution of $\|\mathbf{X}_t\|$ is algebraically complicated.) Thus there is no good reason to expect, *a priori*, that Gaussian techniques should work in handling $\sup_t \|\mathbf{X}_t\|$. Nevertheless, the same general approach does to work (at least in the case $T = \Re^k$ and i.i.d. components – see, for example, the treatment of χ^2 processes in Leadbetter, Lindgren and Rootzén (1983) and of χ^2 fields Adler (1981)) but the extension is not trivial and involves considerable hard work.

One way to get results while remaining in a purely Gaussian framework is to rewrite the Euclidean norm in (1.59) as an operator type norm: i.e.

(1.60) $$\|\mathbf{X}_t\| := \sup_{\mathbf{a}\in S^N} \mathbf{a}\cdot\mathbf{X}_t,$$

where $S^N = \{\mathbf{a}\in\Re^N : \|\mathbf{a}\| = 1\}$ is the unit sphere in $\Re^N$.

If we now introduce a new, real valued Gaussian process $X(\mathbf{a},t)$ on $S^N \times T$ by the correspondence

(1.61) $$X(\mathbf{a},t) = \mathbf{a}\cdot\mathbf{X}_t,$$

then

(1.62) $$\sup_{t\in T}\|\mathbf{X}_t\| \equiv \sup_{(\mathbf{a},t)\in S^N\times T} X(\mathbf{a},t).$$

The righthand supremum here is of a real valued Gaussian process, albeit over a somewhat involved parameter space, and so in principle is amenable to the theory that we shall develop to handle such suprema in the following chapters.

In fact, it is easy to relate the canonical metric of $X(\mathbf{a},t)$ to that of X_t, since

$$\begin{aligned} E\big(X(\mathbf{a},s) - X(\mathbf{b},t)\big)^2 &= E\big(\mathbf{a}\cdot\mathbf{X}_s - \mathbf{b}\cdot\mathbf{X}_t\big)^2 \\ &= E\big(\mathbf{a}\cdot(\mathbf{X}_s - \mathbf{X}_t) + (\mathbf{a}-\mathbf{b})\cdot\mathbf{X}_t\big)^2. \end{aligned}$$

Consider the special case of i.i.d. components with $\sigma_T^2 = \sup_T EX_t^2$. Then the above gives us

$$\begin{aligned} d^2\big((\mathbf{a},s),(\mathbf{b},t)\big) &\le \|\mathbf{a}\|^2 d^2(s,t) + \|\mathbf{a}-\mathbf{b}\|^2\sigma_T^2 \\ &\qquad + 2\,\mathbf{a}\cdot(\mathbf{a}-\mathbf{b})\,E(X_s^i - X_t^i)X_t^i \\ &\le \|\mathbf{a}\|^2 d^2(s,t) + \|\mathbf{a}-\mathbf{b}\|^2\sigma_T^2 + 2\,\mathbf{a}\cdot(\mathbf{a}-\mathbf{b})\,d(s,t)\sigma_T, \end{aligned}$$

by Cauchy-Schwartz. Note $\|a\| = 1$ to obtain

$$d((\mathbf{a},s),(\mathbf{b},t)) \le d(s,t) + \|\mathbf{a}-\mathbf{b}\|\,\sigma_T .$$

It now follows from simple geometry on the N-dimensional sphere that if we write $N_{S^N,T}(\epsilon)$ for the entropy function of $X(\mathbf{a},t)$, $\mathbf{a} \in S^N$, $t \in T$, and $N_T(\epsilon)$ for that of each of the i.i.d. components X_t^i on T then

$$N_{S^N,T}(\epsilon) \le O\Big(\big(N_T(\epsilon)/\epsilon\big)^N\Big).$$

In the light of (1.62) this is all we need in order to handle those aspects of the distribution of $\sup_{t\in T}\|X_t\|$ that are amenable to entropy methods. ■

If the X_t^i are neither independent nor identically distributed then the above calculation may not be quite as simple, and this approach not as amenable. Nevertheless, this is often a good trick to remember. As we shall see in the next example, it also generalises to much more complex situations.

However, before we leave vector valued processes, it is worthwhile to note that a number of interesting questions arise for these processes, involving delicate interplays between the dimension N of the state space, the dimensionality of the parameter space, and measures of smoothness of the covariance function, that either do not arise, or are trivial, in the context of real valued processes.

For example, suppose the parameter space is $\Re^k$, the component processes X_i are independent, and for each $i = 1,\dots,N$,

$$(1.63)\qquad E|X_s^i - X_t^i|^2 \asymp \|t-s\|^{2\beta_i} \qquad \text{as } \|t-s\| \to 0,$$

where of necessity $0 < \beta_i \le 1$ for all i and we write $f(t) \asymp g(t)$ as $t \to t_0$ if there exist finite constants c and C such that

$$(1.64)\qquad c \le \lim_{t\to t_0} \frac{f(t)}{g(t)} \le C.$$

A process of this form is called a (k,N) Gaussian field of index $\beta = (\beta_1,\dots,\beta_N)$.

Let $\dim A$ denote the Hausdorff dimension of a Euclidean set A. Then, rearranging indices, if necessary, so that $0 < \beta_1 \le \dots \le \beta_N \le 1$,

(1.65)

$$\dim(\operatorname{Im}\mathbf{X}) = \min\left[N,\ \frac{k+\sum_{i=1}^N(\beta_N-\beta_i)}{\beta_N}\right],$$

(1.66)

$$\dim(\operatorname{Gr}\mathbf{X}) = \min\left[k+\sum_{i=1}^N(1-\beta_i),\ \frac{k+\sum_{i=1}^N(\beta_N-\beta_i)}{\beta_N}\right],$$

(1.67)

$$\dim(\mathbf{X}^{-1}(\mathbf{u})) = \max\left[0,\ k-\sum_{i=1}^N\beta_i\right],$$

where (1.65) and (1.66) hold with probability one, and (1.67) holds with positive probability for each $\mathbf{u} \in \Re^k$.

For more details on results of this kind, including a formal definition of Hausdorff dimension, see, for example, Chapter 8 of Adler (1981).

7. BANACH SPACE VALUED PROCESSES:

Banach space valued Gaussian processes arise in a number of different contexts, the two most common being in the areas of empirical processes and stochastic partial differential equations (SPDE's). Here I only want to show you that the basic questions of continuity and boundedness of these processes can be handled, if one so desires, in the framework of real valued processes. The argument is much the same as that we just saw for vector valued processes.

Thus, let B be a separable Banach space, and B^* its topological dual. A stochastic process X_t taking values in B is called Gaussian if for every $n \geq 1$, and all collections $t_1, \ldots, t_n \in T$ and $\phi_1, \ldots, \phi_n \in B^*$ the n-dimensional random variable

$$\big(\langle \phi_1, X_{t_1} \rangle, \ldots, \langle \phi_n, X_{t_n} \rangle\big)$$

has a multivariate Gaussian distribution. The covariance functional will be of the form

$$R_{st}(\phi, \psi) = E \langle \phi, X_s \rangle \langle \psi, X_t \rangle. \tag{1.68}$$

The continuity question for these processes is generally two-fold. Firstly, it is generally required that for fixed $t \in T$, $X_t(.)$ be a continuous linear functional on B^*. Secondly, for fixed $\phi \in B^*$, one wants that $\langle \phi, X_t \rangle$ be continuous in t. Since both of these problems relate to real valued processes, in the first case indexed by B^* and in the second by T, both fall within the realm of the general theory we have espoused so far.

If the continuity question is one of joint continuity in t and ϕ, then the approach developed in the previous example suggests defining the new (real valued) process

$$X(\phi, t) = \langle \phi, X_t \rangle,$$

with covariance function

$$R\big((\phi, s), (\psi, t)\big) = R_{st}(\phi, \psi),$$

and studying its continuity on $B^* \times T$.

This approach is of particular interest in the theory of SPDE's. There one is often interested in an equation which has a solution only in an appropriate family of functions (i.e. Banach space B) which must be specified in advance. In this case arguments of the kind used in the treatment of generalised Gaussian fields above often help one to limit, or at least to guess, the appropriate solution space in advance.

To be fair we should note that the trick of simplifying the state space at the cost of complicating the parameter space is not always easy to apply in practice, since entropy or similar calculations have to be made over $B^* \times T$, which will not generally be a nice space to work with. The most common problem is that T and B^* are very different kinds of spaces. An example of the application of this idea can be found in the paper by Iscoe *et. al.* (1989) in which the continuity of ℓ^2-valued Ornstein-Uhlenbeck processes is treated.

Before leaving this family of examples, however, it is worth noting that Banach space valued Gaussian processes are often not difficult to deal with directly, and, much as in the way that the geometric structure of the parameter space T does not enter the general theory of Gaussian processes, neither (usually) does the structure of the state space. Thus, many of the central results of these notes hold, and are almost as easy to prove, for Banach space instead of real valued processes. (There are, of course, too many annoying exceptions to this rule.) I chose to work with the real valued case in general – perhaps so that the "Introduction" part of the title of the notes will have some validity. Three recent papers with vector-valued results are Fernique (1987a,b, 1988).

8. NON-GAUSSIAN PROCESSES: A natural question to ask is whether or not the results and methods presented in these notes extend naturally to non-Gaussian stochastic processes. The answer, in general, must be somewhat equivocal.

In some cases there is an immediate transfer of results. To take a trivial but often useful example, for $F: \Re \to \Re$ and X_t Gaussian, the process $Y_t = F(X_t)$ is generally non-Gaussian, but continuity questions on Y are clearly simple, given enough information on X and F; i.e. the continuity question for the non-Gaussian Y can be reduced to a question on the Gaussian X. There are some questions, however, that even in this setting have surprising answers. For example, we shall see in the following chapter that a basic inequality that plays a central rôle in the theory of Gaussian processes (Slepian's inequality) fails to work for as simple a process as $|X_t|$.

A similar, but more interesting situation arises in the study of the so-called χ^2-processes. Given a sequence $X_1(t), \ldots, X_N(t)$ of independent copies of a Gaussian process, the corresponding χ^2-process is defined as

$$Z(t) = \sum_{i=1}^{N} X_i^2(t). \tag{1.69}$$

Again, it is clear that the continuity and boundedness questions for the non-Gaussian Z are equivalent to those for the Gaussian X_i, and no new methodolgy needs to be introduced. The comments of the preceeding paragraph, however, indicate that, *a fortiori*, Slepian's inequality cannot hold in this case either.

What if the process X_t, defined, as usual, on a metric space (T,τ), has no relationship whatsoever to a Gaussian process? (Note, that having left the Gaussian framework, we have also left the canonical metric d out of the picture.) Then Pisier (1980, 1981) and Fernique (1983) have various versions the following result, for which you need to know that a function $\phi\colon \Re \to \Re$ is called a Young function if it is even, continuous, convex, and satisfies

$$\lim_{x\to 0} \frac{\phi(x)}{x} = 0, \qquad \lim_{x\to\infty} \frac{\phi(x)}{x} = \infty.$$

1.12 THEOREM. *Let X_t be a Banach space valued stochastic process, defined on a metric space (T,τ), and assume that the process $\|X_t - X_s\|$ is separable. Let N_τ be the metric entropy function for T with respect to the metric τ. If there exists an $\alpha \in (0,1]$ and a Young function ϕ such that the following two conditions are satisfied, then X is continuous with probability one.*

$$E\phi\left(\frac{\|X(t)-X(s)\|^\alpha}{\tau(s,t)}\right) \leq 1, \tag{1.70}$$

$$\int_{N_\tau(u)>1} \phi^{-1}\left(N_\tau(u)\right) du < \infty. \tag{1.71}$$

The power of this result is clear. One immediate implication is that much of the hard work that we have done in calculating entropy functions for Gaussian processes can be used with little or no extra effort to study the continuity properties of non-Gaussian processes as well. The only technical difficulty lies in finding a ϕ satisfying both (1.70) and (1.71). We shall look at only one, albeit very rich, class of examples – the so-called ξ-radial processes of Marcus (1987).

Recall that a symmetric, real valued, infinitely divisible random variable ξ is determined by a characteristic function of the form

$$Ee^{i\lambda\xi} = e^{-\Psi(|\lambda|)}, \tag{1.72}$$

where

$$\Psi(|\lambda|) = \int_0^\infty (\cos\lambda t - 1)\, d\nu[t,\infty), \tag{1.73}$$

and ν is a Lévy measure; i.e. a positive measure on $\Re_+ \setminus \{0\}$ satisfying

$$\int_0^\infty (1\wedge t^2)\, d\nu[t,\infty) < \infty.$$

When $\Psi(|\lambda|) = |\lambda|^\alpha$, then ξ is symmetric stable of index $\alpha \in (0, 2]$. (The case $\alpha = 2$ is, of course, the Gaussian situation, and is somewhat different to the other cases. Nevertheless, we are certainly not lacking for a theory in this case.)

A real valued stochastic process X_t on T is called ξ-radial if there exists a probability measure m on the space of real valued functions $\beta(t)$ for which $\sup_{t\in T} |\beta(t)| = 1$, equipped with the cylindrical σ-algebra, such that for all measurable functions $\gamma(t)$

$$(1.74)\quad E\Big\{\exp\Big(\sum_{t\in T}\gamma(t)X(t)\Big)\Big\} = \exp\Big\{-\int\Psi\Big(\Big|\sum_{t\in T}\gamma(t)\beta(t)\Big|\Big)\,m(d\beta)\Big\}.$$

These processes are strictly stationary (why?), and the measure m is generally referred to as the spectral measure of X. Despite its name, however, it should not be confused with the usual spectral measure of a stationary stochastic process.

ξ-radial processes provide an interesting generalisation of Gaussian processes, and have, in one form or another, been of considerable interest over the past decade. In particular, while they are often highly non-Gaussian in their sample path behaviour, they are nevertheless amenable to a similar style of mathematical analysis. The stable case has been of particular interest, and you can find a good survey of these processes in Weron (1984). The forthcoming monograph of Samorodnitsky and Taqqu (199?) will provide the most wide ranging survey, however, along with many interesting and useful examples of stable processes beyond and including the ξ-radial case. Two of the simplest examples are given by the stable moving averages and harmonizable processes on $\Re$, representable, respectively, by integrals of the form

$$(1.75)\qquad \int_{\Re} H(t-s)\,Z(ds), \qquad \int_{\Re} e^{it\lambda}\,Z(dy),$$

where H is a nice function and Z a stable measure on $\Re$. (i.e. a stable process with independent increments.)

In the stable case it is not too hard to obtain the following corollary of Theorem 1.12. The proof (e.g. Fernique (1983)) relies, however, on various inequalities for stable variables and processes that I do not really want to develop here.

1.13 COROLLARY. *Let X_t be a real valued, symmetric stable process on (T, τ) of index $\alpha \in (0, 1]$, and for $\beta \in (0, \alpha)$ let d_β be the distance function on T defined by*

$$(1.76)\qquad d_\beta(s,t) = E|X_t - X_s|^\beta .$$

Let N_β be the entropy function for (T, d_β). If, for some $\beta \in (0, \alpha)$

$$\int_0 \left(N_\beta\left(\epsilon^{1/\alpha}\right)\right)^{\beta/\alpha} \epsilon^{\beta-1}\, d\epsilon \;<\; \infty, \tag{1.77}$$

then X has continuous sample paths on T with probability one.

The paper of Marcus and Pisier (1984) and the more technical monograph of Marcus (1987) provide the deepest results on the general continuity problem for strictly stationary stable processes, and I shall do no more now than to refer you there to see not only how to improve on Theorem 1.12 for ξ-radial processes, but also to see how to go part of the way towards obtaining necessary, as well as sufficient, conditions for continuty.

It is worthwile noting at this stage, however, that whereas we shall soon see that for Gaussian processes it is possible to develop general necessary and sufficient conditions for sample path continuity, this is not a simple project for ξ-radial processes. Talagrand (1988b, 1989) has provided some necessary but not sufficient conditions for the boundedness and continuity of symmetric stable processes – in terms of majorising measures (see Chapter 4) – but this seems to be a problem whose solution is beyond our current means. (This is primarily because of the lack of one very specific tool – Slepian's inequality – that does not hold in the ξ-radial case but is extremely important in the necessity arguments in the Gaussian case.)

Finally, it is interesting to note that the discussion of Section 2 above, on the Brownian family of processes defined via Gaussian white noise, can be extended to function and set indexed processes defined via infinitely divisible white noise. In this case the resulting processes are rarely continuous, but often have cadlag versions (right continuous with left limits, in an appropriate sense). As was the case for continuity, entropy arguments can also be used to determine cadlaguity. For details see Adler and Feigin (1984) and Bass and Pyke (1984).

4. Exercises.

SECTION 1.1:

1.1 Show that any stochastic process has mean square continuous sample paths if and only if its covariance function R is continuous on $T \times T$. Check that this is equivalent to $\lim_{u \to 0} \rho(u) = 0$, where ρ is the function defined by (1.1). Show also that if R is continuous at diagonal points (t,t), then it is continuous everywhere.

SECTION 1.2:

2.1 Let W be a k-dimensional ($k > 1$) Brownian sheet, and let $(i_1, \ldots, i_n)$, $n < k$, be a subset of $(1, 2, \ldots, k)$. Fix $s_{i_1}, \ldots, s_{i_n}$, $s_{i_j} \geq 0$ for all j, and let

Z be the process on $\Re_+^{k-n}$ defined by

$$Z(t_1,\dots,t_{k-n}) = W(t_1,\dots,t_{i_1-1},s_{i_1},t_{i_1+1}\dots,t_{i_n-1},s_{i_n},t_{i_n+1},\dots,t_{k-n}).$$

i.e. n of the k parameters of W are fixed by the s_{i_j}. Show that

$$|s_{i_1}\cdots s_{i_n}|^{-1/2}\, Z(t_1,\dots,t_{k-n})$$

is a $(k-n)$-parameter Brownian sheet.

2.2 Let $\mathcal{A}_\gamma$ be the collection of subsets of $\Re^2$ described in §1.2 and whose construction is illustrated in Figure 1.2. Using an argument similar to that used to show that the Brownian sheet is unbounded on lower layers, show that it is also unbounded on $\mathcal{A}_\gamma$ if $1<\gamma\le 2$.

SECTION 1.3:

3.1 Let X be a continuous Gaussian process on $[0,1]$, f a homeomorphism of $[0,1]$, and Y a new process defined, also on $[0,1]$, by $Y_t = X_{f(t)}$. Show that X and Y have identical entropy functions. Show, by example, that it is possible for the covariance function of X to satisfy (1.14) while that of Y does not.

3.2 Show that the question of convergence or divergence of the integral (1.14) is equivalent to that of the integral

$$\int_K^\infty p(e^{-x^2})\,dx,$$

for arbitrary $K>0$. Hence, or otherwise, show that (1.15) implies (1.14).

3.3 Assume for the moment that for a stationary Gaussian process on $\Re^1$ (1.14) is both necessary and sufficient for sample path continuity. Using the equivalence established in the previous exercise, show how the continuity/discontinuity question is answered by the relationship (1.18). The assumption will be justified in Chapter 4.

3.4 Prove the easy half of Theorem 1.6 concerning the Talagrand expansion. Let $\{Y_n\}_{n\ge 0}$ be a centered Gaussian sequence satisfying (1.28), and $\{\alpha_n\}_{n\ge 0}$ a sequence of functions satisfying (1.29) on a compact metric space T.
(i) Show that the process X_t defined by (1.30) is bounded with probability one. (You will need the inequality (2.1) for the proof, so use it freely for now. If you want a hint, see the argument following Theorem 4.6, which shows that entropy conditions don't always give tight conditions for continuity, and which therefore motivates half of these notes.)
(ii) It follows from the calculations used to prove (i) and the results of Chapter 4 that X is also continuous, but since we don't know these results

yet, show that the a.s. continuity of X will follow if we assume, in addition to the above, that the $\alpha_n(t)$ are continuous in t.

3.5 Show that if X_t is a centered Gaussian field on a compact subset of $\Re^k$, $k \geq 1$, and $p(u)$ is defined by (1.31), then the convergence of the entropy integral at (1.14) is still a sufficient condition for the continuity of X. Note that the proof for general k is really no different to that for $k=1$.

3.6 This is an exercise for the true lover of entropy methods: Theorem 1.7 gave sufficient conditions for the continuity of a Gaussian process indexed by a family of functions. Show that if the inequality in (1.34) also holds in the opposite direction (obviously with a different C) then the entropy calculations in the proof of Theorem 1.7 are sharp: i.e. the upper bound (1.48) also serves (obviously with different constants) as a lower bound. While this does not immediately imply that the conditions of the theorem are also necessary, it will give us some nice results when we come to talking about extrema distributions in Chapter 5.
(Hint: The route you will have to follow can be found on page 311 of Kolmogorov and Tihomirov (1959), which is the basic reference for all calculations of this kind.)

3.7 Complete the proof of Proposition 1.8.

3.8 Show that the finite union of Vapnik-Červonenkis classes is again a Vapnik-Červonenkis class.

3.9 Let $\mathcal{P}(k,n)$ be the set of n-sided polygons in $\Re^k$. Show that $\mathcal{P}(k,n)$ is a Vapnik-Červonenkis class of sets, and find its VC index.

3.10 If $\mathcal{F}$ is a family of functions on a space E, then for each $\phi \in \mathcal{F}$ set $\text{pos}(\phi) = \{t \in E: \phi(t) > 0\}$ and $\text{pos}(\mathcal{F}) = \{\text{pos}(\phi): \phi \in \mathcal{F}\}$. If $\mathcal{F}$ is a k-dimensional real vector space, such as polynomials of order $\leq k$, show that $\text{pos}(\mathcal{F})$ is a Vapnik-Červonenkis class with VC index k.

II. TWO BASIC RESULTS

Despite the title of this Chapter, there are probably *three* basic results in the theory of Gaussian processes, that make this theory both manageable and special. The first is the existence theorem that to any positive semi-definite function R there corresponds a centered Gaussian process with covariance function R; an important, but not particularly exciting result.

The second is that the supremum of a Gaussian process behaves much like a single Gaussian variable with variance equal to the largest variance achieved by the entire process. In the way that we shall present it, this is Borell's inequality, and is the key to all results about Gaussian continuity, boundedness, and suprema.

The third is that if two centered processes have identical variances (i.e. $EX_t^2 = EY_t^2$ for all $t \in T$), but one process is more "correlated" than the other (i.e. if $EX_tX_s \geq EY_tY_s$ for all $s, t \in T$) then the more correlated process has the stochastically smaller maximum, in the sense that $P\{\sup X_t > \lambda\} \leq P\{\sup Y_t > \lambda\}$ for all $\lambda > 0$. This is Slepian's inequality, and without this result many of the most basic results in the theory of Gaussian processes would have no proof.

Both Borell's and Slepian's inequality are very special in that analagous results for non-Gaussian processes are extremely rare. (We shall see some exceptions to this rule later). The fact that even for Gaussian processes the $\sup X_t$ in Slepian's inequality cannot be replaced by as simple a variant as $\sup |X_t|$ is also indicative how very lucky we are that a result of this kind holds at all.

1. Borell's Inequality.

Let X be a centered Gaussian random variable with variance σ^2. Then choosing

$$\Psi(\lambda) = (2\pi)^{-\frac{1}{2}} \int_\lambda^\infty e^{-\frac{1}{2}x^2}\, dx,$$

to denote the standard Gaussian distribution function, straightforward approximations give that for all $\lambda > 0$

$$(2.1) \qquad \begin{aligned} (1-\sigma^2\lambda^{-2})(\sigma/\sqrt{2\pi})\lambda^{-1}e^{-\frac{1}{2}\lambda^2/\sigma^2} &\leq P\{X > \lambda\} \\ &= \Psi(\lambda/\sigma) \\ &\leq (\sigma/\sqrt{2\pi})\lambda^{-1}e^{-\frac{1}{2}\lambda^2/\sigma^2} \end{aligned}$$

One immediate consequence of (2.1) is that

$$(2.2) \qquad \lim_{\lambda\to\infty} \lambda^{-2} \log P\{X > \lambda\} = -(2\sigma^2)^{-1}.$$

There is a classical result of Landau and Shepp (1970) and Marcus and Shepp (1971) that gives a result closely related to (2.2), but for the supremum of a general centered Gaussian process. If we assume that $\{X_t\}_{t\in T}$ has bounded sample paths with probability one, then they showed that

$$\lim_{\lambda\to\infty} \lambda^{-2} \log P\{\sup_{t\in T} X_t > \lambda\} = -(2\sigma_T^2)^{-1}, \tag{2.3}$$

where

$$\sigma_T^2 := \sup_{t\in T} EX_t^2$$

is a notation that will remain with us for the remainder of these notes. An immediate consequence of (2.3) is that for all $\epsilon > 0$ and large enough λ

$$P\{\sup_{t\in T} X_t > \lambda\} \leq e^{\epsilon\lambda^2 - \frac{1}{2}\lambda^2/\sigma_T^2}. \tag{2.4}$$

Since $\epsilon > 0$ is arbitrary, comparing (2.4) and (2.1) we reach the conclusion described above that the supremum of a centered, bounded Gaussian process behaves much like a single Gaussian variable with a suitably chosen variance.

In Chapter 5 we shall investigate (2.4) in considerable detail, and show how to close the gap between (2.4) and (2.1) (i.e. between λ^{-1} and $e^{\epsilon\lambda^2}$).

Most proofs of results like (2.3) rely on geometrical arguments and the so-called Brunn-Minkowski inequality for Gauss space (k-dimensional Euclidean space with a k-dimensional Gaussian measure). The strongest form is due to Borell (1975) in a highly abstract setting and with a difficult proof. Maurey and Pisier (Pisier (1986)) recently found a very short proof of a version of Borell's inequality, which avoids the need to appeal to areas outside of probability theory. This is, in essence, the proof that we shall give. It has the advantage of being more self-contained for a probabilistic audience, and the disadvantage that it cannot reach all the cases that proofs based on isoperimetric inequalities can. Nevertheless, it is my favourite application of Itô's formula, for who would have expected to be able to use stochastic analysis to prove results in Gaussian processes? (By the way – the stochastic analysis/Gaussian process interface is now a two way street. See Chapter 6 for details on this.) The result is:

2.1 THEOREM. *Let $\{X_t\}_{t\in T}$ be a centered Gaussian process with sample paths bounded* a.s. *Let* $\|X\| = \sup_{t\in T} X_t$. *Then* $E\|X\| < \infty$, *and for all* $\lambda > 0$

$$P\{|\,\|X\| - E\|X\|\,| > \lambda\} \leq 2e^{-\frac{1}{2}\lambda^2/\sigma_T^2}. \tag{2.5}$$

An immediate consequence of (2.5) is that for all $\lambda > E\|X\|$,

$$P\{\|X\| > \lambda\} \leq 2e^{-\frac{1}{2}(\lambda - E\|X\|)^2/\sigma_T^2}. \tag{2.6}$$

Thus (2.3) and (2.4) are easily seen to be consequences of Borell's inequality.

Indeed, a far stronger result is true, for (2.4) can be replaced by

$$(2.7) \qquad P\{\sup_{t\in T} X_t > \lambda\} \;\le\; e^{C\lambda - \frac{1}{2}\lambda^2/\sigma_T^2},$$

where C is a constant depending on $E\|X\|$.

Of course, the sharper forms (2.5) and (2.6) will only be useful if we can manage to calculate $E\|X\|$. This, in fact, is one of the main tasks facing us, and we shall see that this single expectation is the key to a Pandora's box of other results.

Theorem 2.1 is true in much more generality than we have indicated, and can also be formulated somewhat differently. Borell's original result, for example, used the median of $\|X\|$ instead of the mean $E\|X\|$ in (2.5). In this formulation, the process X can be allowed to take values in a quite general Banach space, and $\|\ \|$ is then the norm of the Banach space. (In fact, since the passage from the inequality with the median to that with the mean, or *vice versa*, is far from immediate, it is really not quite precise to refer to (2.5) as "Borell's" inequality. Nevertheless, we shall not let a minor point like this change our nomenclature.)

Similar results, involving Banach space valued processes, using both the natural norm and its expectation, are also available, but with a constant other than $\frac{1}{2}$ in the exponent in (2.5). (For details see Pisier (1986, 1989).)

Throughout these notes you should always remember that $\|\ \| \equiv \sup$ is not a true norm, and that very often one needs bounds on the tail of $\sup_t |X_t|$ rather than $\|X\| = \sup_t X_t$. However, a symmetry argument immediately gives one that

$$P\{\sup_t |X_t| > \lambda\} \;\le\; 2P\{\sup_t X_t > \lambda\},$$

so that Borell's inequality helps out here as well.

For more on the relation between stochastic analysis and isoperimetric inequalities, see, for example, Ledoux (1988) and Pisier (1986, 1989). ■

The following lemma forms the main step in the proof of Borell's inequality, and is also of considerable independent interest. (As usual, we shall also denote the usual Euclidean norm by $\|\ \|$, hopefully without creating too much confusion.)

2.2 LEMMA. *Let $f\colon \Re^k \to \Re$ have derivatives of up to second order, bounded pointwise by $Ae^{B\|x\|}$ for some $A, B < \infty$, and let X be a k-dimensional, centered, Gaussian variable with covariance matrix V. If $|f(x) - f(y)| \le \|x - y\|$ for all $x, y \in \Re^k$, then for all $\lambda > 0$*

$$(2.8) \qquad P\{|f(X) - Ef(X)| > \lambda\} \;\le\; 2e^{-\frac{1}{2}\lambda^2/\sigma^2},$$

where

$$\sigma^2 = \sup_{1\le i\le k} V(i,i) = \sup_{1\le i\le k} EX_i^2.$$

REMARK: It may seem strange at first that the upper bound depends neither on the function f nor the covariance structure of the X_i (other than via the variances). If one rewrites the result in the form of (2.6), however, the fact that $Ef(X)$ now appears in the bound, and the fact that this expectation depends explicitly on f and implicitly on the full covariance structure of the X_i, shows that all is in heuristic order.

PROOF: Let $\{B_t\}_{t\ge 0} = \{B_t^1, ..., B_t^k\}_{t\ge 0}$ be a k-dimensional Brownian motion; i.e. the B^i are i.i.d. standard, real-valued Brownian motions.

We can link the vector X to B_t by noting that $X_1,\dots,X_k$ is distributed exactly like $V^{\frac12}B_1$, where $V^{\frac12}$ satisfies $V = V^{\frac12}\cdot(V^{\frac12})'$, and exists by virtue of the positive semi-definiteness of V. Before we can fully utilise this fact, however, we need to make two excursions into stochastic analysis.

The first starts by noting that if $G\colon \Re^k \to \Re^k$ is continuous and coordinatewise bounded, and if $\langle\ ,\ \rangle$ is the usual Euclidean inner product then

$$\exp\left\{\int_0^t \langle G(B_s), dB_s\rangle - \tfrac12\int_0^t \|G(B_s)\|^2\,dt\right\}$$

is an (exponential) martingale with initial value, and so constant mean, 1. (For information on exponential martingales, or, indeed, on any of the stochastic analysis arguments that follow, Karatzas and Shreve (1988) is a very accessible reference. In this case, the requisite result is on page 199.) Taking expectations, and setting

$$\alpha = \sup_{x\in\Re^d} \|G(x)\|,$$

we obtain that for all real θ

$$E\left\{\exp\left(\theta\int_0^1 \langle G(B_s), dB_s\rangle\right)\right\} \le e^{\frac12\theta^2\alpha^2}.$$

A standard Chebycheff type argument gives us that

$$\begin{aligned}
&P\left\{\left|\int_0^1 \langle G(B_s), dB_s\rangle\right| > \lambda\right\}\\
&\quad= P\left\{\int_0^1 \langle G(B_s), dB_s\rangle > \lambda\right\} + P\left\{\int_0^1 \langle G(B_s), dB_s\rangle < -\lambda\right\}\\
&\quad\le 2e^{-\theta\lambda}\, E\left\{\exp\left(\theta\int_0^1 \langle G(B_s), dB_s\rangle\right)\right\} \qquad (2.9)\\
&\quad\le 2e^{-\theta\lambda}\, e^{\frac12\theta^2\alpha^2}\\
&\quad= 2e^{-\frac12\lambda^2/\alpha^2},
\end{aligned}$$

the factor of two in the first inequality coming from symmetry considerations and the last inequality being a consequence of setting $\theta = \lambda/\alpha^2$.

Our second excursion involves Itô's formula for real valued functions of vector valued Brownian motion. (Karatzas and Shreve (1988), page 153.) The form we shall need states that for a sufficiently smooth $F = F(x,t): \Re^k \times \Re_+ \to \Re$,

$$\begin{aligned} F(B_t,t) - F(B_s,s) &= \int_s^t \langle \nabla_x F(B_u,u), dB_u \rangle \\ &+ \int_s^t \left(\tfrac{1}{2}\Delta_{xx} F(B_u,u) + F_t(B_u,u)\right) du, \end{aligned} \tag{2.10}$$

where ∇_x and Δ_{xx} denote derivatives of $F(x,t)$ with respect to x, and $F_t(x,t) = \partial F(x,t)/\partial t$. We shall also need $\{P_t\}_{t\geq 0}$, the Markov semi-group associated with B, determined by the fact that for smooth $g: \Re^k \to \Re$

$$\begin{aligned} (P_t g)(x) &= E^x g(B_t) \\ &= (2\pi t)^{-k/2} \int_{\Re^k} g(y) e^{-\frac{1}{2}\|x-y\|^2/t}\, dy, \end{aligned}$$

where E^x denotes expectation with respect to the Brownian motion B starting at the point $x \in \Re^k$ at time zero.

We can now put the above two parts together to prove our Lemma.

Let $\hat{f}: \Re^k \to \Re^1$ satisfy the differentiability requirements of the f of the lemma, and assume $|\hat{f}(x) - \hat{f}(y)| \leq \sigma\|x - y\|$. Setting $F(x,t) = (P_{1-t}\hat{f})(x)$, the conditions of the lemma imply that F is sufficiently smooth for Itô's formula to hold. With $t = 1$, $s = 0$, (2.10) yields

$$\hat{f}(B_1) - E\hat{f}(B_1) = \int_0^1 \langle \nabla(P_{1-u}\hat{f})(B_u), dB_u \rangle. \tag{2.11}$$

The last expression in the Itô formula disappears due to the specific form of the semi-group P_t. If you are not familiar with this (it is the heat equation that makes everything work) you should do the algebra to convince yourself that everything works as claimed.

Since P_t is a *contraction* semi-group, the fact that $|\hat{f}(x) - \hat{f}(y)| < \sigma\|x - y\|$ immediately implies that $P_t\hat{f}$ satisfies the same inequality for every $t \geq 0$, and so $\|\nabla P_t \hat{f}(x)\| \leq \sigma$ for almost every x. It then easily follows from (2.9) that

$$P\{|\hat{f}(B_1) - E\hat{f}(B_1)| > \lambda\} \leq 2e^{-\frac{1}{2}\lambda^2/\sigma^2}.$$

To complete the proof note simply that $f(X) \stackrel{\mathcal{L}}{=} f(V^{\frac{1}{2}}B_1)$, so that (2.7) follows from (2.11) with $\hat{f}(x) = f(V^{\frac{1}{2}}x)$. The function $\hat{f}$ then satisfies all the requirements placed on it. ∎

PROOF OF THEOREM 2.1: We have two things to prove. Firstly, Theorem 2.1 would follow immediately from Lemma 2.2 in the case of finite T if only sup(.) were a sufficiently smooth function. Unfortunately, it is not, being non-differentiable on the diagonal in $\Re^k$. Fortunately, however, it is approximable by smooth functions. Any standard approximation procedure (such as convolution with C^∞ functions) will work, and so this part of the proof is left to you.

The second part of the proof involves lifting the result from finite to general T. This is, almost, an easy exercise in approximation.

For each $n > 0$ let T_n be a finite subset of T such that $T_n \subset T_{n+1}$ and T_n increases to a dense subset of T. By separability,

$$\sup_{t \in T_n} X_t \stackrel{\text{a.s.}}{\to} \sup_{t \in T} X_t,$$

and, since the convergence is monotone, we also have that

$$E \sup_{t \in T_n} X_t \to E \sup_{t \in T} X_t.$$

Since $\sigma^2_{T_n} \to \sigma^2_T < \infty$, (again monotonely) this would be enough to prove the general version of Borell's inequality from the finite T version if only we knew that the one worrisome term, $E \sup_T X_t$, were definitely finite, as claimed in the statement of the Theorem. Thus if we now that the assumed a.s. finiteness of $\|X\|$ implies also the finiteness of its mean, we shall have a complete proof to both parts of the Theorem.

We proceed by contradiction. Thus, assume $E\|X\| = \infty$, and choose $\lambda_o > 0$ such that

$$e^{-\lambda_o^2/\sigma_T^2} \le \tfrac{1}{4}, \qquad P\{\sup_{t \in T} X_t < \lambda_o\} \ge \tfrac{3}{4}.$$

Now choose $n \ge 1$ such that $E\|X\|_{T_n} > 2\lambda_o$, possible since $E\|X\|_{T_n} \to E\|X\|_T = \infty$. Borell's inequality on the finite space T_n then gives

$$\begin{aligned}
\tfrac{1}{2} \ge 2e^{-\lambda_o^2/\sigma_T^2} &\ge 2e^{-\lambda_o^2/\sigma_{T_n}^2} \\
&\ge P\{\big|\, \|X\|_{T_n} - E\|X\|_{T_n} \big| > \lambda_o\} \\
&\ge P\{E\|X\|_{T_n} - \|X\|_T > \lambda_o\} \\
&\ge P\{\|X\|_T < \lambda_o\} \\
&\ge \tfrac{3}{4}.
\end{aligned}$$

This provides the required contradiction, and so we are done. ∎

I cannot overemphasise how important a result Borell's inequality is. For example, an almost immediate consequence of Borell's inequality is that the a.s. finiteness of $\|X\|$ implies that it also has all regular, and some exponential, moments. (c.f. Theorem 3.2.) In later chapters, especially Chapter 5, we shall see how one can apply Borell's inequality a number of times, with almost no other tools, to obtain even sharper bounds on tail probabilities for suprema.

Now, however, we turn to the second, and equally central, result about Gaussian processes.

2. Slepian's Inequality.

There are a variety of different ways to present Slepian-like inequalities today. We choose the following formulation, (from Joag-Dev, Perlman and Pitt (1983)) which actually includes a number of interesting side results.

2.3 THEOREM. *Let $X_1, ..., X_k$ be centered Gaussian variables with covariance matrix $R = (r_{ij})_{i,j=1}^k$, $r_{ij} = EX_iX_j$. Let $h: \Re^k \to \Re$ be C^2, and assume that, together with its derivatives, it satisfies a $O(\|x\|^N)$ growth condition at infinity for some finite N. Let*

$$\mathcal{H}(R) = Eh(X_1, \ldots, X_k). \tag{2.12}$$

and assume that for a pair (i,j), $1 \le i < j \le k$

$$\frac{\partial^2 h(x)}{\partial x_i \partial x_j} \ge 0 \qquad \text{for all } x. \tag{2.13}$$

Then $\mathcal{H}(R)$ is an increasing function of r_{ij}.

Note that the theorem is actually true without the growth conditions on h, if one is prepared to attribute veracity to the result if $\mathcal{H}$ is identically infinite. The proof is reasonably straightforward, and, in its important details, goes back to Slepian (1962).

PROOF: We have to show that

$$\frac{\partial \mathcal{H}(R)}{\partial r_{ij}} \ge 0$$

whenever $\partial^2 h / \partial x_i \partial x_j \ge 0$.

To make our lives a little easier we assume that R is non-singular, so that it makes sense to write $\phi(x) = \phi_R(x)$ for the centered Gaussian density on $\Re^k$ with covariance matrix R. Then some algebra (dating back at least to Plackett (1954)) shows that

$$\frac{\partial \phi}{\partial r_{ii}} = \frac{1}{2}\frac{\partial^2 \phi}{\partial x_i^2}, \qquad \frac{\partial \phi}{\partial r_{ij}} = \frac{\partial^2 \phi}{\partial x_i \partial x_j}, \qquad i \ne j. \tag{2.14}$$

(The algebra of this calculation is almost identical to that needed to justify (2.11) in the proof of Borell's inequality!) Apply this result and our assumptions on h to justify two integrations by parts to obtain

$$\frac{\partial \mathcal{H}(R)}{\partial r_{ij}} = \int_{\Re^k} h(x)\frac{\partial \phi(x)}{\partial r_{ij}}dx = \int_{\Re^k} \frac{\partial^2 h(x)}{\partial x_i \partial x_j}\phi(x)dx \geq 0.$$

(If the integrations by parts bother you, see Berman (1987).) This completes the proof for the case of non-singular R. The general case can be handled by approximating a singular R via a sequence of non-singular covariance matrices. ■

Here are some of the consequences of Theorem 2.3:

2.4 COROLLARY (SLEPIAN'S INEQUALITY). *If X and Y are a.s. bounded, centered Gaussian processes on T such that $EX_t^2 = EY_t^2$ for all $t \in T$ and*

$$E(X_t - X_s)^2 \leq E(Y_t - Y_s)^2 \qquad \text{for all } s,t \in T, \tag{2.15}$$

then for all real λ

$$P\{\sup_{t\in T} X_t > \lambda\} \leq P\{\sup_{t\in T} Y_t > \lambda\} \tag{2.16}$$

PROOF: By separability, and the final argument in the proof of Borell's inequality, it suffices to prove (2.16) for T finite. Note that since $EX_t^2 = EY_t^2$ for all $t \in T$, (2.15) implies that $EX_sX_t \geq EY_sY_t$ for all $s,t \in T$. Let $h(x) = \prod_{i=1}^k f_i(x_i)$, where each f_i is a positive non-increasing, C^2 function satisfying the growth conditions placed on h in the statement of Theorem 2.3 and k is the number of points in T. Note that, for $i \neq j$

$$\frac{\partial^2 h(x)}{\partial x_i \partial x_j} = f_i'(x_i)f_j'(x_j)\prod_{\substack{n\neq i\\ n\neq j}} f_n(x_n) \geq 0,$$

since both f_i' and f_j' are non-positive. It therefore follows from the theorem that

$$E\prod_{i=1}^k f_i(X_i) \geq E\prod_{i=1}^k f_i(Y_i).$$

Now take $\{f_i^{(n)}\}_{n=1}^\infty$ to be a sequence of positive, non-increasing, C^2 approximations to the indicator function of the interval $(-\infty, \lambda]$, to derive that

$$P\{\|X\| < \lambda\} \geq P\{\|Y\| < \lambda\}.$$

That is,

$$P\{\|X\| > \lambda\} \;\le\; P\{\|Y\| > \lambda\},$$

which is what we had to prove. ■

Slepian's inequality will form one the basic building blocks of the results of Chapter 4. In general, it is written with the condition $EX_sX_t \ge EY_sY_t$ rather than (2.15), since this more clearly indicates the difference between the dependence structures of the two processes.

It is immediate from symmetry considerations that a similar result holds for infima rather than suprema, as it does for the range of X_t, but, as we shall soon see by example, no such result holds for $\sup_t |X_t|$.

Another consequence of Slepian's inequality is

2.5 COROLLARY. *Under the conditions of Corollary 2.4*

$$E \sup_{t\in T} X_t \;\le\; E \sup_{t\in T} Y_t \tag{2.17}$$

PROOF:

$$\begin{aligned} E\|X\| &= \int_0^\infty P\{\|X\| > \lambda\}\, d\lambda - \int_{-\infty}^0 P\{\|X\| < \lambda\}\, d\lambda \\ &\le \int_0^\infty P\{\|Y\| > \lambda\}\, d\lambda - \int_{-\infty}^0 P\{\|Y\| < \lambda\}\, d\lambda \\ &= E\|Y\|. \end{aligned}$$

This proves the result. ■

The main problem with Slepian's inequality is that it requires that X_t and Y_t have identical variances. To compare processes with differing variances it turns out that we have to concentrate on inequalities of the form (2.17) rather that (2.16). Our first result in this regard is the almost trivial

2.6 OBSERVATION. *If X is an a.s. bounded, centered Gaussian process on T and Y a centered Gaussian variable (not necessarily independent of X), then*

$$E\Big(\sup_{t\in T}(X_t + Y)\Big) \;=\; E \sup_{t\in T} X_t \tag{2.18}$$

As trite as this observation is, it ceases to be valid if, for example, we investigate $\sup_t |X_t|$ rather than $\sup_t X_t$. It has surprisingly important implications. For example, it allows us to prove

2.7 LEMMA. *Let X and Y be two centered Gaussian random variables. Then*

$$E\max(X,Y) = \frac{\left(E(X-Y)^2\right)^{\frac{1}{2}}}{\sqrt{2\pi}}. \tag{2.19}$$

PROOF: Set $\sigma^2 = E(X-Y)^2$. By Observation 2.6

$$\begin{aligned} E\max(X,Y) &= E\max(X-Y, Y-Y) \\ &= E\max(X-Y, 0) \\ &= \frac{1}{\sqrt{2\pi}\sigma}\int_0^\infty z\, e^{-z^2/2\sigma^2}\, dz \\ &= \frac{\sigma}{\sqrt{2\pi}}, \end{aligned}$$

as required. ■

A useful inequality that follows from this lemma is that for any a.s. bounded, centered X,

$$E\sup_{t\in T} X_t \geq \frac{\left(\sup_{s,t\in T} E(X_s - X_t)^2\right)^{\frac{1}{2}}}{\sqrt{2\pi}}. \tag{2.20}$$

What is more important, however, is that Observation 2.6 will allow us to prove

2.8 THEOREM. *If X and Y are a.s. bounded, centered Gaussian processes on T such that*

$$E(X_t - X_s)^2 \leq E(Y_t - Y_s)^2 \tag{2.21}$$

then

$$E\sup_T X_t \leq 2E\sup_T Y_t. \tag{2.22}$$

Note that the important difference between Lemma 2.5 and Theorem 2.8 is not so much the additional factor of 2 on the right hand side, but the fact that we no longer require that X and Y have identical variances.

PROOF: Fix a point $t_o \in T$ and set $\alpha^2 = \sup_{t\in T} E(Y_t - Y_{t_o})^2$. Furthermore, set

$$\widetilde{X}_t = X_t - X_{t_o}, \qquad \widetilde{Y}_t = Y_t - Y_{t_o},$$

and let η and η' be two standard Gaussian variables, independent of one another and the two processes X_t and Y_t. Now define

$$\widehat{X}_t = \widetilde{X}_t + \eta\alpha,$$
$$\widehat{Y}_t = \widetilde{Y}_t + \eta' g(t),$$

where

$$g^2(t) := \alpha^2 - E(Y_t - Y_{t_0})^2 + E(X_t - X_{t_0})^2.$$

Note that $\alpha^2 \geq g^2(t) \geq 0$ for all $t \in T$, and that $E\widehat{X}_t^2 = E\widehat{Y}_t^2$ for all $t \in T$. Furthermore, since

$$\begin{aligned} E(\widehat{Y}_t - \widehat{Y}_s)^2 &= E[\widetilde{Y}_t - \widetilde{Y}_s) + \eta'(g(t) - g(s))]^2 \\ &= E(\widetilde{Y}_t - \widetilde{Y}_s)^2 + (g(t) - g(s))^2 \\ &\geq E(\widetilde{Y}_t - \widetilde{Y}_s)^2, \end{aligned} \tag{2.23}$$

it follows that

$$\begin{aligned} E(\widehat{X}_t - \widehat{X}_s)^2 &= E(\widetilde{X}_t - \widetilde{X}_s)^2 \\ &\leq E(\widetilde{Y}_t - \widetilde{Y}_s)^2 \\ &\leq E(\widehat{Y}_t - \widehat{Y}_s)^2, \end{aligned}$$

the first inequality following from (2.21) and the second from (2.23).

Observation 2.6 gives us that

$$E\|X\| = E\|\widetilde{X}\| = E\|\widehat{X}\|,$$

and Corollary 2.5 gives us that

$$E\|\widehat{X}\| \leq E\|\widehat{Y}\|.$$

But,

$$\begin{aligned} E\|\widehat{Y}\| &= E\sup_{t\in T}(\widetilde{Y}_t + \eta' g(t)) \\ &\leq E\|\widetilde{Y}\| + E\sup_{t\in T}\eta' g(t) \\ &\leq E\|\widetilde{Y}\| + E\{\sup_{t\in T}\eta' g(t) 1_{\eta'>0}\} \\ &= E\|\widetilde{Y}\| + \sup_{t\in T} g(t)\, E\{\eta' 1_{\eta'>0}\} \\ &\leq E\|\widetilde{Y}\| + \frac{\alpha}{\sqrt{2\pi}}, \end{aligned}$$

the last line following from the definition of g and a standard Gaussian integration. However, by (2.20), the last line here is less that $2E\|\widetilde{Y}\| = 2E\|Y\|$. This concludes the proof. ■

There is, in fact, a stronger version of Theorem 2.8, in which the factor of 2 in (2.22) does not appear. The proof is somewhat more involved than that of Theorem 2.8, and relies on new calculations rather than a clever application of previous results.

2.9 THEOREM (SUDAKOV-FERNIQUE INEQUALITY). *Under the conditions of Theorem 2.8, (2.22) continues to hold if the factor of 2 is removed from the right hand side of the inequality.*

PROOF: Actually, I shall only show you how to start the proof, and shall leave the last part, which involves considerable calculus, up to you. You can find the details in Fernique (1975) or Jain and Marcus (1978).

The main trick of the proof lies in noting that it is sufficient to show that

$$E \sup_{s,t\in T} |X_t - X_s| \leq E \sup_{s,t\in T} |Y_t - Y_s|.$$

The result we seek will then follow since, for example,

$$\begin{aligned} E \sup_{s,t\in T} |Y_t - Y_s| &= E \sup_{s,t\in T} (Y_t - Y_s) \\ &= E\{\sup_{t\in T} Y_t + \sup_{t\in T}(-Y_t)\} \\ &= 2\,E\|Y\|. \end{aligned}$$

As usual, we assume that T is finite with k points, and use separability and the final argument in the proof of Borell's inequality to complete the proof. Let $\mathbf{X}$ and $\mathbf{Y}$ denote the k-vectors of values of X_t and Y_t on T. Put copies of $\mathbf{X}$ and $\mathbf{Y}$ onto the same probability space, and assume they are independent. For $\theta \in [0,1]$ set

$$\mathbf{Z}(\theta) = \sqrt{1-\theta}\,\mathbf{X} + \sqrt{\theta}\,\mathbf{Y},$$

and $\psi(\theta) = E(\sup_i (Z(\theta))_i)$. We need to show that $\psi(0) \leq \psi(1)$, for which it is sufficient to show that $\psi'(\theta) \geq 0$ for all $\theta \in [0,1]$.

If $\mathbf{R}(\theta)$ is the covariance matrix of $\mathbf{Z}(\theta)$, then using Fourier transforms we have that the k-dimensional density of $\mathbf{Z}(\theta)$ is given by

$$p_\theta(\mathbf{z}) = (2\pi)^{-k} \int_{\Re^k} \exp\left\{i\langle \mathbf{z},\mathbf{x}\rangle - \tfrac{1}{2}\langle \mathbf{R}(\theta)\mathbf{x},\mathbf{x}\rangle\right\} d\mathbf{x}.$$

Since

$$\psi(\theta) = \int_{\Re^k} \max\{z_1,\ldots,z_k\}\, p_\theta(\mathbf{z})\, d\mathbf{z},$$

we have

$$\psi'(\theta) = \int_{\Re^k} \max\{z_1,\ldots,z_k\} \frac{\partial p_\theta(\mathbf{z})}{\partial \theta}\, d\mathbf{z}.$$

If you now start differentiating, keeping (2.14) in mind, you will find that

$$\psi'(\theta) = \frac{1}{2}\sum_{i,j} \frac{\partial r_{ij}(\theta)}{\partial \theta} \int_{\Re^k} \max\{z_1,\ldots,z_k\} \frac{\partial^2 p_\theta(\mathbf{z})}{\partial z_i \partial z_j}\, d\mathbf{z}.$$

What remains to complete the proof is to show that this expression is positive. This is left to you, with the help of the references given above. ■

The Sudakov-Fernique inequality is a good place to look to see how far one can generalize results of this kind. For example, the inequality cannot work for $\sup |X_t|$. To see this, let Z be a centered Gaussian variable, ξ a positive real number, and X a bounded Gaussian process. Let Y^ξ be the process defined by

$$Y_t^\xi = \xi Z + X_t, \qquad t \in T.$$

Then $E(Y_s^\xi - Y_t^\xi)^2 = E(X_s - X_t)^2$ for all $s,t \in T$, but $E\sup_{t\in T} |Y_t^\xi| \to \infty$ as $\xi \to \infty$, while $E\sup_{t\in T} |X_t|$ is bounded.

To show that the original Slepian's inequality also fails for $\sup_{t\in T} |X_t|$, take $T=\{1,2\}$, with X_1 and X_2 standard normal with correlation ρ. Writing $P_\rho(\lambda)$ for the probability under correlation ρ that $\max(X_1,X_2) > \lambda$ and Ψ as usual for the right hand tail of the standard normal distribution function we see that

$$P_{-1}(\lambda) = P_{-1}\{\max(X_1, X_2) > \lambda\} = P\{|X| > \lambda\} = 2\Psi(\lambda)$$

$$P_0(\lambda) = 2\Psi(\lambda) - \Psi^2(\lambda)$$

$$P_{+1}(\lambda) = P_1\{\max(X_1, X_2) > \lambda\} = P\{X_1 > \lambda\} = \Psi(\lambda)$$

Hence $P_{-1}(\lambda) \geq P_0(\lambda) \geq P_1(\lambda)$ as Slepian's inequality requires. But if $\hat{P}_\rho$ is the probability that $\max(|X_1|,|X_2|) > \lambda$, then $\hat{P}_{-1}(\lambda) = \hat{P}_{+1}(\lambda) = 2\Psi(\lambda)$, $\hat{P}_0(\lambda) = 4[\Psi(\lambda) - \Psi^2(\lambda)]$, so that for all $\lambda > 0$

$$\hat{P}_{-1}(\lambda) < \hat{P}_0(\lambda) \qquad \hat{P}_0(\lambda) > \hat{P}_1(\lambda),$$

and the monotonicity required by Slepian breaks down. ■

Here is a list of interesting variations of Slepian's inequality. You can skip them without harming your understanding of the following chapters, but it would be a shame.

ELLIPTICALLY CONTOURED DISTRIBUTIONS: If $p_\Sigma(x)$ is a density on $\Re^k$ of the form

$$p_\Sigma(x) = |\Sigma|^{-\frac{1}{2}} p(x'\Sigma^{-1}x),$$

where Σ is an invertible, positive definite matrix and $p\colon [0,\infty) \to \Re_+$ is assumed to satisfy $\int_0^\infty \lambda^{k-1} p(\lambda^2)\, d\lambda < \infty$, we say that p is *elliptically contoured.* A version of Slepian's inequality holds for such densities, for writing $\Sigma = (\sigma^{ij})_{1\le i,j\le n}$ and defining

$$\mathcal{H}(\Sigma) = \int_{\Re^k} h(x) p_\Sigma(x) dx,$$

with h satisfying the appropriate continuity and growth conditions (as in Theorem 2.3), then Joag-Dev *et al.* have shown that $\mathcal{H}(\Sigma)$ is an increasing function of σ^{ij}. The proof is almost identical to that we gave for Theorem 2.3.

GORDON'S INEQUALITY: An interesting extension of Slepian's inequality, which is a result about Gaussian maxima, is a result of Gordon's about the min-max of a rectangular array of Gaussian variables. Let $(X_{ij})_I$, $(Y_{ij})_I$, $I = \{(i,j)\colon 1 \le i \le n, 1 \le j \le m\}$ be two collections of centered Gaussian variables satisfying the following three conditions:

$$\begin{aligned} EX_{ij}^2 &= EY_{ij}^2 && (i,j) \in I,\\ EX_{ij}X_{ik} &\le EY_{ij}Y_{ik} && (i,j),(i,k) \in I,\\ EX_{ij}X_{\ell k} &\ge EY_{ij}Y_{\ell k} && (i,j),(\ell,k) \in I,\ i \ne \ell. \end{aligned}$$

Then, for all real λ_{ij},

$$P\Big\{\bigcap_{i=1}^n \bigcup_{j=1}^m [X_{ij} > \lambda_{ij}]\Big\} \ge P\Big\{\bigcap_{i=1}^n \bigcup_{j=1}^m [Y_{ij} > \lambda_{ij}]\Big\}.$$

This implies, for example, that for any increasing function g on $\Re$

$$E\{\min_{1\le i\le n} \max_{1\le j\le m} g(X_{ij})\} \ge E\{\min_{1\le i\le n} \max_{1\le j\le m} g(Y_{ij})\}, \tag{2.24}$$

and that for all $\lambda > 0$

$$P\{\min_{1\le i\le n} \max_{1\le j\le m} X_{ij} \ge \lambda\} \ge P\{\min_{1\le i\le n} \max_{1\le j\le m} Y_{ij} > \lambda\}. \tag{2.25}$$

This result has extensions from minmax to minmaxmin..., etc. and to elliptically contoured distributions as well. It is a consequence of the following theorem, whose proof is due to Kahane (1986).

2.10 THEOREM. *Let* $\mathbf{X} = (X_i)$ *and* $\mathbf{Y} = (Y_i)$, $i = 1,\dots,k$ *be two collections of Gaussian variables, and* I *and* J *subsets of* $\{1,\dots,k\}^2$ *such that*

$$EX_iX_j \le EY_iY_j \qquad (i,j) \in I, \tag{2.26}$$

$$EX_iX_j \ge EY_iY_j \qquad (i,j) \in J, \tag{2.27}$$

$$EX_iX_j = EY_iY_j \qquad (i,j) \notin I \cup J. \tag{2.28}$$

Let $h\colon \Re^k \to \Re$ be C^2, and assume that, together with its derivatives, it satisfies a $O(\|x\|^N)$ growth condition at infinity for some finite N. Furthermore, assume

$$\frac{\partial^2 h(x)}{\partial x_i \partial x_j} \geq 0 \qquad (i,j) \in I, \tag{2.29}$$

$$\frac{\partial^2 h(x)}{\partial x_i \partial x_j} \leq 0 \qquad (i,j) \in J, \tag{2.30}$$

then

$$Eh(\mathbf{X}) \leq Eh(\mathbf{Y}). \tag{2.31}$$

Note that Theorem 2.3, and so Slepian's inequality and all that it implied, are consequences of this result. Thus what follows is, in fact, an alternative proof of Slepian's inequality.

PROOF: As in the proof of Theorem 2.9, put copies of $\mathbf{X}$ and $\mathbf{Y}$ onto the same probability space, and assume they are independent. For $t \in [0,1]$ set

$$\mathbf{Z}(\theta) = \sqrt{1-\theta}\,\mathbf{X} + \sqrt{\theta}\,\mathbf{Y},$$

and $\psi(\theta) = Eh(\mathbf{Z}(\theta))$. We need to show that $\psi(0) \leq \psi(1)$, for which it is sufficient to show that $\psi'(\theta) \geq 0$ for all $\theta \in [0,1]$. Writing h_i for $\partial h(x)/\partial x_i$, and allowing ourselves the luxury of interchanging the order of expectation and differentiation, we have

$$\psi'(\theta) = \sum_{j=1}^{N} E\{h_j(\mathbf{Z}(\theta))\, Z_j'(\theta)\}. \tag{2.32}$$

Fix θ and j, and note that

$$E\{Z_i(\theta)\, Z_j'(\theta)\} = \tfrac{1}{2} E\{Y_j Y_i - X_j X_i\}.$$

It thus follows from (2.26)–(2.28) that if $W_1, \ldots, W_N$ is a new sequence of Gaussian variables, independent of both the X_i and Y_i, then we can express the Z_i as follows:

$$Z_i(\theta) = \alpha_i Z_j'(\theta) + W_i, \tag{2.33}$$

where $\alpha_i \geq 0$ if $(i,j) \in I$, $\alpha_i \leq 0$ if $(i,j) \in J$, and $\alpha_i = 0$ if $(i,j) \notin I \cup J$.

With j still fixed, consider the behaviour of a typical summand in (2.33). Then

(2.34)
$$E\{h_j(\mathbf{Z}(\theta))Z_j'(\theta)\} = \int \cdots \int h_j(\alpha_1 z_j' + w_1, \ldots, \alpha_N z_j' + w_N) z_j' \times \phi_{Z_j'}(z_j')\phi_W(w_1, \ldots, w_N)\, dz_j' dw_1 \ldots dw_N,$$

where $\phi_{Z_j'}$ and ϕ_W are the obvious Gaussian densities.

To complete the argument, differentiate (2.34) with respect to each α_i, and note, that from the properties of the α_i (c.f. the line following (2.33)) and the inequalities (2.29) and (2.30), that the resulting expression is positive if $(i,j) \in I$, and negative if $(i,j) \in J$. That is, $E\{h_j(\mathbf{Z}(\theta))Z_j'(\theta)\}$ is an increasing function of α_i for $(i,j) \in I$, and decreasing for $(i,j) \in J$.

On the other hand, it is clear that $E\{h_j(\mathbf{Z}(\theta))Z_j'(\theta)\} = 0$ if all the $\alpha_i = 0$. Consequently, this expectation must always be non-negative. That is, $\psi'(\theta) \geq 0$, which is what we had to prove. ■

As one might expect, Gordon's inequality can be extended in much the same way as Fernique's inequality can. For example, there is an analagous result for elliptically contoured distributions. You can find further results, as well as applications of Gordon's inequality, (which we shall return to in the following section) in Gordon (1987, 1988a,b). Note also from the proof of Theorem 2.10 that the restriction to twice differentiable h is unnecessarily restrictive. It would have been enough, for example, to assume that the first order derivatives h_j are all absolutely continuous with positive generalised derivatives.

INFINITELY DIVISIBLE PROCESSES: We consider only a simple example.

Let μ be a non-negative measure on $\Re^d$ and let N_μ be a Poisson point process on $\Re^d$ with intensity μ. Let $A_1, \ldots, A_n$ be Borel sets in $\Re^d$ and define $X_i = N_\mu(A_i)$. Let $Y_i = N_\nu(A)$ be similarly defined with respect to a Poisson process with intensity ν. If, for all $B \subseteq \{1, \ldots, n\}$, $B \neq \emptyset$,

(2.35)
$$\mu\Big(\bigcap_{i \in B} A_i\Big) \leq \nu\Big(\bigcap_{i \in B} A_i\Big), \qquad \mu\Big(\bigcup_{i \in B} A_i\Big) \geq \nu\Big(\bigcup_{i \in B} A_i\Big),$$

then for all $\lambda > 0$

(2.36)
$$P\{\max_{1 \leq i \leq n} X_i > \lambda\} \geq P\{\max_{1 \leq i \leq n} Y_i > \lambda\}.$$

That is, a form of Slepian's inequality holds. Since Poisson processes are the natural building blocks of infinitely divisible processes, it is not hard to see that a result of this form must extend to a far more general situation. For details, see Brown and Rinott (1988) and Ellis (1988).

The primary interest in (2.36) is that it allows one to build multidimensional Kolmogorov-Smirnov tests based on multivariate empirical distribution functions, which, unlike the examples discussed earlier, hold for finite sample size. You can find details (and tables!) in Adler, Brown and Lu (1988).

The fact that a Slepian-like inequality holds for these processes is somewhat surprising, since there is nothing Gaussian-like in their structure. Thus it is *not* surprising that whereas the proofs of the basic Gaussian result, the results for elliptically contained distributions, and the min-max result all look basically the same, the proof of (2.36) under (2.35) is completely different.

STABLE PROCESSES: We have already discussed these processes in the previous Chapter. It has long been part of the folklore of stable processes that no Slepian like inequality holds, and counter-examples have been constructed to demonstrate this: e.g. Fernique (1983), and Linde (1986). In a certain sense, however, this claim is not quite true. The examples given all show that ordering two stable processes in terms of the *covariation* function, does not imply stochastic ordering of the sample maxima. (The covariation function is a stable analogue of the covariance function for Gaussian processes, and is defined as

$$\operatorname{cov}(s,t) \;=\; \int \beta(s)|\beta(t)|^{\alpha-1}\operatorname{sgn}(\beta(t))\, m(d\beta),$$

where m is the spectral measure of (1.74).)

However, whereas covariance functions determine Gaussian processes, the same is not true of covariation functions for stable processes. Thus, it is not necessarily natural to try to base a stochastic majoration result on comparison of covariation functions. Gennady Samorodnitsky has recently shown me a Slepian type result for stable processes based on comparison of spectral measures. Hopefully it will be written up sometime soon.

One factor to note, however, is that one of the reasons that Slepian's inequality is so useful is that it is so very easy to state and apply. This is not the case with Samorodnitsky's result for stable processes. Thus it still seems that a "pure" Slepian inequality is basically a Gaussian result only, and we are very lucky to have it.

DIFFUSIONS AND THEIR ILK: There is a substantial body of results in the diffusion literature on stochastic domination problems. For example, let $X_1(t)$ and $X_2(t)$ be two diffusions on $[0,\infty)$ satisfying the stochastic differential equations

$$dX_i(t) \;=\; a_i\big(X_i(t)\big)\,dt + b\big(X_i(t)\big)\,dW_i(t), \tag{2.37}$$

where the $W_i(t)$ are standard Brownian motions. Note that the X_i have different drift, but the same "speed" function. If $a_1(x) \le a_2(x)$ for all x, then

$$P\{\|X_1\|_T > \lambda\} \;\le\; P\{\|X_2\|_T > \lambda\}$$

for all λ and all $T \subset [0,\infty)$. Indeed, much more than this Slepian like inequality is true, for one can build X_1 and X_2 on the same sample space in such a way that, with probability one, $X_1(t) \leq X_2(t)$ for all $t \in [0,\infty)$. The construction is easy. Just start the processes off in such a way that $X_1(0) \leq X_2(0)$, and use the same Brownian motion in (2.37) to generate each of the two processes. The fact that $a_1(X) \leq a_2(X)$ will then ensure that $X_1(t) \leq X_2(t)$ for all t.

In the light of this result, it is natural to ask the following question: Given two Gaussian processes satisfying the conditions of Slepian's inequality, can one construct them on the same sample space so that the maximum of one a.s. dominates that of the other? Simple examples with two-point parameter spaces show that this is not the case.

Once again, to convince you that in spite of what we have just indicated Slepian's inequality is a primarily Gaussian result, you should remember that the diffusion result is heavily restricted to processes on the real line only, and has no extension (as, in fact, can be said of diffusions themselves) to unstructured parameter spaces.

3. Applications in Banach Spaces.

One of the most fruitful areas in the theory of probability over the past decade has been the development of an active interface with the theory of Banach spaces. Much of this activity has centered around the general theory of Gaussian processes, and in this section I want to briefly describe two closely related results in the local theory of Banach spaces and their proof via Gaussian methods.

It is beyond the intended scope of these notes to enter into a detailed treatment of these results. Nevertheless, I do want you to see them, if only to show how useful the general theory of Gaussian processes is, even outside the usual settings of probability and stochastic processes. For details you should see the notes by Pisier (1986) and his brand new monograph Pisier (1989), as well as the monographs by Milman and Schechtman (1986) and Linde (1986). These cover not only results of the kind discussed below, but also a wide variety of other applications of Gaussian processes to Banach space theory.

We start by recalling some basic facts and definitions. Let E, F be Banach spaces and let $\lambda > 1$. Then E and F are called λ-isomorphic if there is an isomorphism $T\colon E \to F$ such that $\|T\| \cdot \|T^{-1}\| \leq \lambda$, where, as usual,

$$\|T\| = \sup_{\|x\|_E \leq 1} \|T(x)\|_F ,$$

and $\| \ \|_E$ and $\| \ \|_F$ and the norms on E and F. If E and F are λ-isomorphic we write $E \overset{\lambda}{\sim} F$.

This concept can be used to define a (Banach-Mazur) distance between E and F by

$$\rho(E,F) = \inf\{\lambda : E \overset{\lambda}{\sim} F\}.$$

The importance of this distance function is that it is a good measure of how close two Banach spaces are in terms of their local characteristics. To provide an example of this, let ℓ_p^n denote the Euclidean space $\Re^n$ equipped with the norm $\|x\|_p = (\sum_{i=1}^n |x_i|^p)^{1/p}$. Here is a result dating back, in essence, to Dvoretzky (1961):

2.11 THEOREM. *Let B be a Banach space, and $\epsilon > 0$. Then there exists a $n = n(\epsilon)$ such that B has a subspace F with $F \overset{1+\epsilon}{\sim} \ell_2^n$.*

Here is a more precise result in a more specific setting.

2.12 THEOREM. *Let $\epsilon > 0$ and $n \geq 1$. If $N \geq \alpha \exp(\beta n \epsilon^{-2})$ is an integer (where $\alpha, \beta > 0$ are universal constants), then for every convex set $B \subset \Re^N$ which contains the origin in its interior, there is a subspace F of dimension n and a constant $a > 0$, such that*

$$aS^n \subseteq F \cap B \subseteq \left(\frac{1+\epsilon}{1-\epsilon}\right) aS^n, \tag{2.38}$$

where S^n is the unit ball in $\Re^n$.

These two results are really versions of the same basic truth, Theorem 2.12 being the sharper of the two because of the additional information it provides on n. The bound $\alpha \exp(\beta n \epsilon^{-2})$ that it gives can actually be shown to be sharp.

Neither of these results looks like it has anything to do with Gaussian processes, but here is the briefest of outlines of a proof that will show there is a very strong connection indeed.

Recall firstly the comment made after the statement of Theorem 2.1 that Borell's inequality (2.5) holds also for Banach space valued Gaussian processes in which the norm in (2.5) is the Banach norm.

To construct the subspace F of Theorem 2.11, let X be a B-valued Gaussian variable, take $n \geq 1$, and let $X_1, \ldots, X_n$ be independent copies of X. Let $M = E\|X\|$. Then some reasonably simple calculations, put together with the Banach space version of Borell's inequality, show that for the right values of n there is a subset Ω_0 of our generic probability space, with $P\{\Omega_0\} > 0$, such that for $\omega \in \Omega_0$ the following is true for all $x \in \Re^n$:

$$(1+\epsilon)^{-\frac{1}{2}}\left(\sum_{i=1}^n |x_i|^2\right)^{\frac{1}{2}} \leq M^{-1}\left\|\sum_{i=1}^n x_i X_i(\omega)\right\| \leq (1+\epsilon)^{\frac{1}{2}}\left(\sum_{i=1}^n |x_i|^2\right)^{\frac{1}{2}}.$$

Taking the span of $\{X_1(\omega), \ldots, X_n(\omega)\}$ for $\omega \in \Omega_0$ provides an example of F.

This proves Theorem 2.11. Details can be found in Pisier (1986, 1989). The proof of Theorem 2.12 follows the same principle; i.e. one uses random Gaussian mappings satisfying certain conditions with positive probability to provide proof that such mappings exist at all. The proof of Theorem 2.12, however, is based on a very clever use of Gordon's inequality (2.24). The proof is in Gordon (1988).

4. Exercises.

SECTION 2.1:

1.1 By expanding the normal probability density in a power series, or otherwise, establish the two sides of the basic inequality (2.1).

1.2 If you have not already done so, prove (2.11) by applying Itô's formula (2.10) to $F(x,t) = (P_{1-t}\hat{f})(x)$. Note how very neatly various complicated terms cancel out, and marvel in the beauty and speciality of the Gaussian density which makes this happen.

SECTION 2.2:

2.1 Show by example on a two point parameter space that the a.s. domination of suprema that holds for diffusions does not hold for Gaussian processes. Why does it not hold for Gaussian diffusions?

III. PRELUDE TO CONTINUITY

1. Boundedness and Continuity.

The principle aim of this section is to show that for Gaussian processes the question of sample path continuity is intimately related to the boundedness of the supremum. In one direction this is obvious and, indeed, non-probabilistic. If the parameter space T is compact, then the a.s. continuity of X implies the a.s. boundedness of $\sup_{t \in T} |X_t|$. Thus the problem is essentially to find conditions under which processes with bounded suprema are also continuous.

Recall that we treat only centered processes, and measure continuity in terms of the canonical metric $d(s,t) = (E(X_s - X_t)^2)^{1/2}$ on T.

The first result we need is the following easy lemma, which tells us that as far as a.s. boundedness is concerned it is irrelevant whether we work with $\sup X_t$ or $\sup |X_t|$.

3.1 LEMMA. *For X centered, Gaussian, on T, and $t_o \in T$*

$$E \sup_{t \in T} X_t \;\leq\; E \sup_{t \in T} |X_t| \;\leq\; E|X_{t_o}| + 2E \sup_{t \in T} X_t.$$

PROOF: Only the rightmost inequality needs proving. Note the trivial inequalities that for any $t, t_o \in T$

$$X_t - X_{t_o} \;<\; \sup_{t \in T}(X_t - X_{t_o}), \qquad X_{t_o} - X_t \;<\; -\inf_{t \in T}(X_t - X_{t_o}).$$

Furthermore, both $\sup_{t \in T}(X_t - X_{t_o}) \geq 0$ and $-\inf_{t \in T}(X_t - X_{t_o}) \geq 0$. Applying the relationships $\max(a, -a) = |a|$ and $\max(a,b) \leq a+b$ if $a, b \geq 0$, it follows from the above two inequalities that

$$\begin{aligned} |X_t| &\leq |X_{t_o}| + |X_t - X_{t_o}| \\ &\leq |X_{t_o}| + \sup_{t \in T}(X_t - X_{t_o}) - \inf_{t \in T}(X_t - X_{t_o}). \end{aligned}$$

Taking a supremum over the left hand side leaves the right side unchanged. Now take expectations, and note that symmetry gives us that $E \sup_T X_t = -E \inf_T X_t$ to complete the proof. ∎

The next result is somewhat more interesting and important, and is a good example of the power of Borell's inequality. The original proof of this result, which appears as Exercise 1.1, is due to Fernique (1978), and involves a comparatively long and sophisticated calculation.

3.2 THEOREM. *For X centered, Gaussian,*

$$(3.1)\qquad P\{\sup_{t\in T} X_t < \infty\} = 1 \iff E\sup_{t\in T} X_t < \infty \iff Ee^{\alpha\|X\|^2} < \infty$$

for sufficiently small α.

REMARK: Theorem 3.2 can be strengthened considerably. For the best possible results, which tell you exactly how large α can be for something like (3.1) to hold, see Talagrand (1984). The proof will not use the fact that $\|\ \|$ is the supremum function, and so the result will hold in the same generality as Borell's inequality itself.

PROOF OF THEOREM 3.2: The existence of the exponential moments of $\|X\|$ implies the existence of $E\|X\|$, and this in turn implies the a.s. finiteness of $\|X\|$. Furthermore, since by Theorem 2.1 we already know that the a.s. finiteness of $\|X\|$ entails that of $E\|X\|$, all that remains is to prove is that the a.s. finiteness of $\|X\|$ also implies the existence of exponential moments.

But this is an easy consequence of Borell's inequality, since, with both $\|X\|$ and $E\|X\|$ now finite,

$$\begin{aligned} Ee^{\alpha\|X\|^2} &= \int_0^\infty P\{e^{\alpha\|X\|^2} > \lambda\}\,d\lambda \\ &= E\|X\| + \int_{E\|X\|}^\infty P\{\|X\| > \sqrt{\log\lambda^{1/\alpha}}\}\,d\lambda \\ &\le E\|X\| \\ &\quad + 2\int_{E\|X\|}^\infty \exp\{-\tfrac{1}{2}(\sqrt{\log\lambda^{1/\alpha}} - E\|X\|)^2/\sigma_T^2\}\,d\lambda \\ &\le E\|X\| \\ &\quad + 4\alpha\int_0^\infty u\exp\{-\tfrac{1}{2}(u - E\|X\|)^2/\sigma_T^2\}\exp\{\alpha u^2\}\,du, \end{aligned}$$

which is clearly finite for small enough α. ∎

The next theorem is the first to relate continuity to boundedness, and, in essence, is the result that will enable us to concentrate only on the expected values of suprema in the future, and derive all other results from results on these.

3.3 THEOREM. *Let X be a.s. bounded on T and let τ be a metric on T such that the canonical metric d is τ-uniformly continuous. Then X is τ-uniformly continuous with probability one if, and only if, $\lim_{\eta\to 0}\phi_\tau(\eta) = 0$, where ϕ_τ is given by*

$$(3.2)\qquad \phi_\tau(\eta) = E\sup_{\tau(s,t)<\eta}(X_s - X_t).$$

PROOF: We start with necessity. For each ω we have

$$\lim_{\eta\to 0}\sup_{\tau(s,t)<\eta}|X_s(\omega)-X_t(\omega)| = 0,$$

so that the fact that $\lim_{\eta\to 0}\phi_\tau(\eta)=0$ follows from dominated convergence (c.f. Theorem 3.2).

Conversely, since d is τ-uniformly continuous, we can find a sequence η_n with $\phi_\tau(\eta_n)\le 2^{-n}$, such that $\tau(s,t)<\eta_n$ implies $d(s,t)<2^{-n}$. Consider the event

$$A_n = \{\sup_{\tau(s,t)<\eta_n}|X_s-X_t| > 2^{-n/2}\}. \tag{3.3}$$

Since X is, by assumption, a.s. bounded, we can apply Borell's inequality (Theorem 2.1) to obtain that, for $n\ge 3$,

$$\begin{aligned} P\{A_n\} &\le 2\exp\big(-\tfrac{1}{2}(2^{-n/2}-2^{-n})^2/2^{-2n}\big) \\ &\le K\exp\big(-2^{n-1}\big). \end{aligned}$$

Since $P\{A_n\}$ is an admirably summable series, Borel-Cantelli gives us that X is a.s. uniformly τ-continuous, as required. This completes the proof. ∎

The proof of Theorem 3.3 actually yields more than what was claimed in the statement, which is why we departed from our regular policy of stating everything in terms of the canonical metric only. In fact, if we denote the τ-modulus of (uniform) continuity of X by

$$W_\tau(\eta) = \sup_{\tau(s,t)<\eta}|X_s-X_t|, \tag{3.4}$$

then the above calculations actually give us substantial information on the size of W_τ. Thus we have that not only is the comparatively simple question of sample path continuity inextricably tied up with the question of boundedness, but much finer information on moduli of continuity comes for free in this formulation.

3.4 COROLLARY. *Assume that the conditions of Theorem 3.3 hold and that* $\lim_{\eta\to 0}\phi_\tau(\eta)=0$. *Then, for all* $\epsilon>0$ *there exists an a.s. finite random variable* $\delta=\delta(\omega)$ *such that, for almost all* ω,

$$W_\tau(\eta) \le \phi_\tau(\eta)|\log\phi_\tau(\eta)|^\epsilon, \tag{3.5}$$

for all $\eta\le\delta(\omega)$. *That is,* $\phi_\tau(\cdot)|\log\phi_\tau(\cdot)|^\epsilon$ *is a uniform sample modulus for* X *in the metric* τ.

REMARK: The ϵ in (3.5) is, as are all such infinitesimals, exceedingly irritating, and one would like to dispose of it. That this can in fact be done will be shown in the following chapter. As you read the following proof, note that if only (3.6) could be replaced with $d_\tau(\eta) = o(\phi_\tau(\eta))$ for small η, then virtually the same proof would suffice to establish the stronger result. In order to obtain such an inequality, however, we have to have sharp tools for calculating $E\|X\|$. But this is precisely what Chapter 4 is about.

PROOF: Set

$$\begin{aligned} d_\tau(\eta) &:= \sup_{\tau(s,t)\le\eta} d(s,t) \\ &= \sup_{\tau(s,t)\le\eta} \left(E|X_t - X_s|^2\right)^{\frac{1}{2}}, \end{aligned}$$

and note the trivial inequality

$$\begin{aligned} d_\tau(\eta) &= \sup_{\tau(s,t)<\eta} \left(E|X_t - X_s|^2\right)^{\frac{1}{2}} \\ &= \sqrt{2\pi} \sup_{\tau(s,t)<\eta} E|X_t - X_s| \\ &\le 2\sqrt{2\pi}\, E \sup_{\tau(s,t)<\eta} (X_t - X_s) \\ &= 2\sqrt{2\pi}\, \phi_\tau(\eta), \end{aligned} \tag{3.6}$$

where the second line is a standard Gaussian result and the inequality follows from Lemma 3.1, applied to the two-parameter process $Y(s,t) = X(s) - X(t)$ with "t_o"$= (t,t)$ for some $t \in T$.

Since, by assumption, ϕ_τ is τ-continuous we can define, for each $n \ge 1$,

$$\eta_n = \inf\{\eta : \phi_\tau(\eta) = e^{-n}\}.$$

Define also

$$B_n = \Big\{ \sup_{\tau(s,t)<\eta_n} |X_s - X_t| > \phi_\tau(\eta_n)|\log\phi_\tau(\eta_n)|^{\epsilon/2} \Big\}.$$

By two (because of the absolute value sign) applications of Borell's inequality,

$$\begin{aligned} P\{B_n\} &\le 4\exp\left\{ -\tfrac{1}{2}\left(|\log\phi_\tau(\eta_n)|^{\epsilon/2} - 1\right)^2 \frac{\phi_\tau^2(\eta_n)}{d_\tau^2(\eta_n)} \right\} \\ &\le K_1 \exp\left\{ -K_2 n^\epsilon \right\} \end{aligned}$$

by (3.6) and the definition of η_n.

Since $\sum_n P\{B_n\} < \infty$, we have that for $n \ge N(\omega)$

$$W_\tau(\eta_n) \le \phi_\tau(\eta_n)|\log\phi_\tau(\eta_n)|^{\epsilon/2}.$$

Monotonicity of W_τ and d, along with separability, complete the proof. ■

Another easy corollary of the proof of Theorem 3.3 is the following.

3.5 COROLLARY. *Let X be as in Theorem 3.3 and for $t \in T$ set*

$$\phi_\tau^t(\eta) = E \sup_{s:\tau(s,t)<\eta} (X_s - X_t).$$

Then X is a.s. τ-continuous at t if, and only if, $\lim_{\eta \to 0} \phi_\tau^t(\eta) = 0$.

The proof is almost verbatim that of the theorem, so we shall not give it. Note that a local modulus of τ-continuity based on ϕ_τ^t instead of ϕ_τ follows exactly as in Corollary 3.4.

At first sight, Corollary 3.5 seems to be of far less interest than Theorem 3.3, since it deals only with continuity at a specific point, rather than throughout T. That this is *not* the case is a consequence of the following unexpected and important result.

2. Zero-One Laws and Continuity.

3.6 THEOREM. *A Gaussian process X on T has continuous sample paths with probability one if, and only if, it is continuous at each fixed $t \in T$ with probability one; i.e.*

$$(3.7) \qquad P\{\lim_{s \to t} X_s = X_t \text{ for all } t \in T\} = 1 \iff P\{\lim_{s \to t} X_s = X_t\} = 1, \text{ for each } t \in T.$$

Necessity is obvious. It is the sufficiency of (3.7) that is the big surprise. There is something very special about Gaussian processes that makes Theorem 3.6 hold. A similar result for, for example, Poisson processes is palpably false.

We shall need a number of preliminary results before we can prove Theorem 3.6. Most of these are of significant interest in their own right and that, in fact, is the main reason for bringing them at this point. However, they, are not necessary for the understanding of the main results of these notes, and so can be skipped at the first reading.

We start with the *reproducing kernel Hilbert space* (RKHS) of a Gaussian process with covariance function R.

In essence, this is made up of functions that have about the same smoothness properties that $R(s,t)$ has, as a function in t for fixed s, or vice versa. Start with

$$S = \left\{ f : T \to \Re : f(\cdot) = \sum_{i=1}^{n} a_i R(s_i, \cdot),\ a_i \text{ real},\ s_i \in T,\ n \geq 1 \right\}.$$

Define an inner product on S by

$$(3.8)\qquad (f,g)_H = \Big(\sum_{i=1}^{n} a_i R(s_i,\cdot), \sum_{j=1}^{m} b_j R(s_j,\cdot)\Big)_H = \sum_{i=1}^{n}\sum_{j=1}^{n} a_i b_j R(s_i,s_j).$$

The fact that R is non-negative definite implies $(f,f)_H \geq 0$ for all $f \in S$. Furthermore, note that the inner product (3.8) has the following unusual property:

$$(3.9)\qquad f(t) = \sum_{i=1}^{n} a_i R(s_i,t) = \Big(\sum_{i=1}^{n} a_i R(s_i,\cdot), R(t,\cdot)\Big)_H = (f, R(t,\cdot))_H .$$

We refer to (3.9) as the "reproducing kernel" property. From this it follows that for $f \in S$, $t \in T$

$$(3.10)\qquad |f(t)|^2 = |(f,R(t,\cdot))_H|^2 \leq (f,f)_H \left(R(t,\cdot),R(t,\cdot)\right)_H ,$$

the inequality being merely the Schwartz inequality for semi-inner products, which holds as long as $(f,f)_H \geq 0$. Thus, if $(f,f) = 0$, then (3.10) implies that $f(t) = 0$ for all $t \in T$. Consequently, (3.8) defines a proper inner product on S, and so we thus obtain a norm $\|f\|_H = (f,f)_H^{1/2}$. For $\{f_n\}_{n\geq 1}$ a sequence in S we have

$$(3.11)\qquad |f_n(t) - f_m(t)|^2 = |(f_n - f_m, R(t,\cdot))_H|^2 \leq \|f_n - f_m\|^2\, \|R(t,\cdot)\|_H^2 \leq \|f_n - f_m\|_H^2\, R(t,t),$$

the last line following directly from (3.8). Thus it follows that if $\{f_n\}$ is Cauchy in $\|\cdot\|_H$ then it is pointwise Cauchy. The closure of S under this norm is a space of real-valued functions, denoted by $H(R)$, and called the RKHS of X or of R, since every $f \in H(R)$ satisfies (3.9) by the separability of $H(R)$. Since T is separable, and R continuous, it follows that $H(R)$ is also separable. In a moment we shall look at two examples of these spaces, and more can be found among the exercises. In general, however, the RKHS is a rather nebulous concept, good more for proving theorems than anything else. The principle exception is the case $T = \Re^1$, in which case the RKHS has been exploited as an important tool in the detection and estimation problems of communication theory. It is also in this case that the RKHS is comparatively easy to identify. ■

For the first example, however, take $T = \{1,\dots,N\}$, finite, and X centered Gaussian with covariance matrix $R = (r_{ij})$. Let $R^{-1} = (r^{ij})$ denote the inverse of R. Then the RKHS of X is made up of all N-dimensional vectors $f = \langle f_1,\dots,f_N\rangle$ with inner product

$$(f,g)_H = \sum_{i=1}^{N}\sum_{j=1}^{N} f_i r^{ij} g_j .$$

To prove this, we need only check that the reproducing kernel property (3.9) holds. But, with $\delta(i,j)$ the Kronecker delta function, and R_k denoting the k-th row of R,

$$\begin{aligned}(f, R_k)_H &= \sum_{i=1}^{N}\sum_{j=1}^{N} f_i r^{ij} r_{kj} \\ &= \sum_{i=1}^{N} f_i \delta(i,k) \\ &= f_k ,\end{aligned}$$

as required. ■

For a slightly more interesting example, take X to be standard Brownian motion on $T = [0,1]$, so that $R(s,t) = \min(s,t)$. Note that the function $R(s,\cdot)$ is differentiable everywhere except at s, so that following the heuristics developed above we expect that $H(R)$ should be made up of a subset of functions that are differentiable almost everywhere.

To both make this statement more precise, and prove it, we start by looking at what the space S looks like. Thus let

$$f(t) = \sum_{i=1}^{n} a_i R(s_i,t), \qquad g(t) = \sum_{i=1}^{n} b_i R(t_i,t)$$

be two elements of S. According to (3.8), the S-inner product between them is given by

$$(f,g)_H = \sum_{i=1}^{n}\sum_{j=1}^{n} a_i b_j \min(s_i,t_j).$$

Note that the derivative of $R(s,t)$ with respect to t is given by $1_{[0,s]}(t)$, where 1_A is the indicator function of $A \subset T$, and so the derivative of f is given by $\sum_{i=1}^{n} a_i 1_{[0,s_i]}(t)$. Therefore, we can rewrite the above as follows:

$$\begin{aligned}
(f,g)_H &= \sum_{i=1}^{n}\sum_{j=1}^{n} a_i b_j \int_0^1 1_{[0,s_i]}(t) 1_{[0,t_j]}(t)\, dt \\
&= \int_0^1 \sum_{i=1}^{n} a_i 1_{[0,s_i]}(t) \sum_{j=1}^{n} b_j 1_{[0,t_j]}(t)\, dt \\
&= \int_0^1 \dot{f}(t)\dot{g}(t)\, dt.
\end{aligned}$$

We can now go about treating the general case. Set

$$H = \left\{ f : f(t) = \int_0^t \dot{f}(s)\, ds, \quad \int_0^1 \left(\dot{f}(s)\right)^2 ds < \infty \right\}, \tag{3.12}$$

and define the following inner product on H:

$$(f,g)_H = \int_0^1 \dot{f}(s)\dot{g}(s)\, ds. \tag{3.13}$$

Since it is immediate that $R(s,\cdot) \in H$ for $t \in [0,1]$, and

$$\left(f, R(t,\cdot)\right)_H = \int_0^1 \dot{f}(s) 1_{[0,t]}(s)\, ds,$$

it now follows that $H = H(R)$. That is, the RKHS is, in this case, determined by the space (3.12) and the inner product (3.13). ■

More examples can be found in the exercises. Now, however, we shall look at the RKHS from a slightly different viewpoint, and see how to use it in a very practical fashion.

Since $H(R)$ is a separable Hilbert space, it must have a countable orthonormal basis. This will be extremely important for us. Sometimes this basis is easy to find, particularly in the case $T = [0,1]^k$, in which case it leads us to the *Karhunen-Loève expansion* of X, a topic we shall cover in the following section. Sometimes, we have to be content with merely knowing it exists. In an abstract setting, however, even this is useful, as we shall soon see.

Define $\mathcal{H}^1(X)$, the so-called "linear part" of the $\mathcal{L}^2$ space of the process X, as the closure in $\mathcal{L}^2(P) = \mathcal{L}^2(\Omega, \mathcal{F}, P)$ of

$$\left\{ \sum_{i=1}^{n} a_i X(t_i), \ a_i \text{ real}, \ t_i \in T, \ n \geq 1 \right\}, \tag{3.14}$$

thinned out by identifying all elements indistinguishable in $\mathcal{L}^2(P)$. (i.e. elements U, V for which $E(U-V)^2 = 0$.) This contains all distinguishable random variables, with finite variance, obtainable as linear combinations of values of the process. There is a linear, one-one mapping between the space S of real-valued functions on T to this $\mathcal{L}^2$ space, defined by

$$\Theta(f) = \Theta\Big(\sum_{i=1}^{n} a_i R(t_i, \cdot)\Big) = \sum_{i=1}^{n} a_i X(t_i).$$

Note that Θ is clearly norm preserving, and so extends to all of $H(R)$ with range equal to all of $\mathcal{H}^1(X)$. The extension is called the *canonical isomorphism* between these spaces.

Since $H(R)$ is separable, we now also know that $\mathcal{H}^1(X)$ is. We can use this to build an orthonormal basis for $\mathcal{H}^1(X)$, for if $\{\phi_n\}_{n\geq 1}$ is an orthonormal basis for $H(R)$, then setting $\xi_n = \Theta(\phi_n)$ gives $\{\xi_n\}_{n\geq 1}$ as an orthonormal basis for $\mathcal{H}^1(X)$. In particular, we must have $E\xi_n = 0$ for all $n \geq 1$, and

$$X_t = \sum_{n=1}^{\infty} \xi_n E(X_t \xi_n), \tag{3.15}$$

where the series converges in $\mathcal{L}^2(P)$. Since Θ was an isometry, it follows from (3.15) that

$$\begin{aligned} EX_t\xi_n &= (R(t,\cdot), \phi_n)_H \\ &= \phi_n(t) \end{aligned} \tag{3.16}$$

the last equality coming from the reproducing kernel property of $H(R)$. Putting (3.16) together with (3.15) is almost enough to give the following central result.

3.7 THEOREM. *If $\{\phi_n\}_{n\geq 1}$ is an orthonormal basis for $H(R)$, then X has the $\mathcal{L}^2$-representation*

$$X_t = \sum_{n=1}^{\infty} \xi_n \phi_n(t), \tag{3.17}$$

where $\{\xi_n\}_{n\geq 1}$ is the orthonormal sequence of centered Gaussian variables given by $\xi_n = \Theta(\phi_n)$.

PROOF: We have proven everything other than the fact that the ξ_n are Gaussian. But this follows from standard properties of Gaussian random variables since a countable, $\mathcal{L}^2$ convergent sum of Gaussian random variables will itself always be Gaussian. ∎

The equivalence in (3.17) is only in $\mathcal{L}^2$; i.e. whereas X_t is defined with probability one, the sum is, in general, convergent only in mean square. The following result is indicative of how much we get for free once we have an a.s. continuous process.

3.8 THEOREM. *If X has continuous sample paths with probability one, then the sum in (3.17) converges uniformly on T with probability one.*

There is also a converse result, that the a.s. uniform convergence of a sum like (3.17) implies the continuity of X. But since we shall soon have better ways of establishing sample function continuity, we have no need of the converse. We shall require one non-standard result from probability theory in order to prove Theorem 3.8. It is due to Itô and Nisio (1968).

3.9 LEMMA. *Let $\{Z_n\}_{n\geq 1}$ be a sequence of symmetric independent random variables, taking values in a separable, real Banach space B, equipped with the norm topology. Let $X_n = \sum_{i=1}^n Z_i$. Then X_n converges with probability one if, and only if, there exists a B-valued random variable X such that $\langle x^*, X_n\rangle \to \langle x^*, X\rangle$ in probability for every $x^* \in B^*$, the topological dual of B.*

We also require another preliminary result, of considerable intrinsic interest.

3.10 LEMMA. *If $\{\phi_n\}_{n\geq 1}$ is an orthonormal basis for $H(R)$, then*

$$\sum_{n=1}^{\infty} \phi_n^2(t)$$

converges uniformly in $t \in T$ to $R(t,t)$.

PROOF: By the orthonormal expansion and the reproducing kernel property,

$$R(t,\cdot) = \sum_{n=1}^{\infty} \phi_n(\cdot)\big(R(t,\cdot),\phi_n\big)_H = \sum_{n=1}^{\infty} \phi_n(\cdot)\phi_n(t), \tag{3.18}$$

convergence of the sum being in the $\|\cdot\|_H$ norm. Hence, $\sum_{n=1}^{\infty}\phi_n^2(t)$ converges to $R(t,t)$ for every $t\in T$. Furthermore, the convergence is monotone, and so it follows that it is also uniform ($\equiv$ Dini's theorem). ■

We can now turn to the

PROOF OF THEOREM 3.8: We know that, for each $t \in T$, $\sum_{n=1}^{\infty} \xi_n \phi_n(t)$ is a sum of independent variables converging in $\mathcal{L}^2(P)$. Thus, by Lemma 3.9, applied to real-valued random variables, it converges with probability one to a limit we denote by X_t. Since the limit process is, by assumption, continuous, this defines the same process as that appearing in the $\mathcal{L}^2$ sense in Theorem 3.7.

Now, consider both X and each function $\xi_n\phi_n(\cdot)$ as elements of the Banach space $C(T)$, with sup-norm topology, and define

$$X_n(\cdot) = \sum_{i=1}^{n} \xi_i\phi_i(\cdot) = \sum_{i=1}^{n} \Theta(\phi_i)\phi_i(\cdot).$$

By Lemma 3.9, it suffices to show that for every $x^* \in C^*(T)$ the random variables $\langle x^*, X_n\rangle$ converge in probability to $\langle x^*, X\rangle$.

Recall that every x^* in the topological dual of $C(T)$ is a finite, signed, Borel measure on T. Thus,

$$\begin{aligned} E|\langle x^*, X_n\rangle - \langle x^*, X\rangle| &= E\left|\int_T (X_n(t) - X(t))x^*(dt)\right| \\ &\leq \int_T E|X_n(t) - X(t)| \cdot |x^*|(dt) \\ &\leq \int_T \left[E(X_n(t) - X(t))^2\right]^{\frac{1}{2}} |x^*|(dt) \\ &= \int_T \Big(\sum_{j=n+1}^{\infty} \phi_j^2(t)\Big)^{\frac{1}{2}} |x^*|(dt), \end{aligned}$$

where $|x^*|(A)$ is the total variation of x^* on $A \subset T$.

Since $\sum_{j=n+1}^{\infty} \phi_j^2(t) \to 0$ uniformly in $t \in T$ by Lemma 3.10, the last expression above tends to zero as $n \to \infty$. Since this implies the convergence in probability of $\langle x^*, S_n\rangle$ to $\langle x, X\rangle$, we are done. ∎

Theorem 3.8 is nice to have, (especially since it comes for free with a.s. continuity) but it is really Theorem 3.7 that is the more important. The fact that X can be written as an infinite sum of independent random variables implies that any property of X that is a "tail event", in that its occurrence or non-occurrence does not depend on "the first n among the ϕ_j", can be expected to have probability zero or one only. This is in fact the case, as we shall see in a moment. Firstly, however, we need a definition.

The *oscillation function* w_X of X is defined as

$$w_X(t) = \lim_{\epsilon\downarrow 0} \sup_{u,v\in B(t,\epsilon)} |X_u - X_v|,$$

where $B(t,\epsilon)$ is the d-ball of radius epsilon centered on $t \in T$. Here is a key result:

3.11 THEOREM. *There exists a (non-random) $\bar{\Re}$-valued upper-semicontinuous function $h(t)$ on T such that*

(3.19) $$P\{w_X(t) = h(t), \text{ for all } t \in T\} = 1.$$

(Recall that $f: T \to \bar{\Re}$ is called upper semi-continuous if $\overline{\lim}_{n\to\infty} f(t_n) = f(t)$ for every sequence $\{t_n\}$ with $\lim_{n\to\infty} t_n = t$.)

Before we give a proof of Theorem 3.11, consider some of the consequences of this result.

PROOF OF THEOREM 3.6: Since h is non-random, the condition of a.s. continuity of X at each point $t \in T$ implies that $h(t) \equiv 0$. Theorem 3.11 allows us to move the "each point" inside the probability statement. ∎

3.12 THEOREM.

$$P\{X \textit{ is continuous for all } t \in T\} = 0 \textit{ or } 1.$$

PROOF: Let h be the function of Theorem 3.11. If $h(t) > 0$ for some t, then, by Theorem 3.11, X is discontinuous at that point, and so $P\{X$ is continuous for all $t \in T\} = 0$. If $h(t) = 0$ for all t, then Theorem 3.11, following the lines of the above proof, also shows that X is continuous with probability one. ∎

The theory of Gaussian processes is rich in such zero-one laws, the most famous perhaps being Belyaev's (1961) dichotomy of stationary process as being either a.s. continuous or a.s. unbounded on every open subset of T, a result which follows from arguments similar to those above. In Exercise 2.3 you can work through a simple proof of a very general zero-one law for vector valued Gaussian variables, that has an elegant extension to processes.

Here is a very general zero-one law for Gaussian processes, that we shall not, however, prove. (c.f. Kallianpur (1969, 1970), Jain (1971) and Cambanis and Rajput, (1973) for details.)

As always, X_t, $t \in T$ is a centered Gaussian process on a probability space $(\Omega, \mathcal{F}, P)$, where $\mathcal{F}$ is assumed to be P-complete. Now let Λ be a space of functions which contains, with probability one, the sample paths of X. Let $\mathcal{B}(\Lambda)$ be the cylindrical σ-algebra of subsets of Λ, define $\Phi: (\Omega, \mathcal{F}, P) \to (\Lambda, \mathcal{B}(\Lambda))$ as

$$\Phi(\omega) = \begin{cases} X(\cdot,\omega) & \text{if } X(\cdot,\omega) \in \Lambda, \\ 0 & \text{if } X(\cdot,\omega) \notin \Lambda. \end{cases}$$

Then Φ clearly induces a probability measure μ on $(\Lambda, \mathcal{B}(\Lambda))$. Let $\bar{\mathcal{B}}(\Lambda)$ be the completion of $\mathcal{B}(\Lambda)$ with respect to μ.

We assume the following two conditions:

(A) Λ is a linear function space under addition of functions and multiplication by scalars.

(B) $H(R) \subset \Lambda$.

3.13 THEOREM. *If (A) and (B) are satisfied, and G is a $\bar{\mathcal{B}}(\Lambda)$-measurable subgroup of Λ, then $\mu(G) = 0$ or 1.*

An immediate corollary of this result is the following:

3.14 COROLLARY. *Let Λ be a set of functions which contains the paths of X, P a.s., and satisifies (A) and (B) above. If $F \in \mathcal{F}$ is such that $F = \Phi^{-1}(G)$, where G is a $\bar{\mathcal{B}}(\Lambda)$-measurable subgroup of Λ, then $P(F) = 0$ or 1.*

You can find a number of applications of this important result in Exercise 2.4. Among them is an alternative proof of Theorem 3.12.

Now, however, we return to a

PROOF OF THEOREM 3.11: Let $B \subseteq T$ be closed, and set

$$w_X(B) = \sup_{s,t \in B} |X_t - X_s|. \tag{3.20}$$

Separability ensures that the $w_X(t)$ of the theorem, as well as $w_X(B)$, are well defined random variables. Let

$$X^{(n)}(t) = \sum_{j=n+1}^{\infty} \xi_j \phi_j(t).$$

Since the ϕ_j are continuous, so must be $\sum_{j=1}^{n} \xi_j \phi_j$, for any $n \geq 1$. It thus follows that for each $w \in \Omega$, $n \geq 1$, and any $B \subset T$,

$$w_{X^{(n)}}(B, w) = w_X(B, w).$$

Since $w_{X^{(n)}}(S)$ is measurable with respect to the σ-algebra generated by $\xi_{n+1}, \xi_{n+2}, \ldots$, it follows by Kolmogorov's zero-one law that, for some fixed number $h = h(B)$,

$$P\{w_X(B, w) = h(B)\} = 1.$$

This defines a set indexed version of h. We still need to exhibit a point indexed version, and to show that it is upper semi-continuous. To this end, let $\mathcal{B}$ be a countable open basis for the topology on T generated by open balls with respect to d, and for $t \in T$ set

$$h(t) = \inf_{B \in \mathcal{B}\,:\, t \in B} h(B).$$

It is immediate from the definition that h is upper-semicontinuous, and not hard to see that

$$\begin{aligned} w_X(t) &= \inf_{B \in \mathcal{B}\,:\, t \in B} w_X(B) \\ &\stackrel{\text{as}}{=} \inf_{B \in \mathcal{B}\,:\, t \in B} h(B) \\ &= h(t), \end{aligned}$$

which is all we need to complete the proof. ■

3. The Karhunen-Loève Expansion.

The Karhunen-Loève expansion of a continuous Gaussian process is really just a special case of the expansion (3.17) when the parameter space T is a compact subset of $\Re^k$. In this case it is possible to choose particularly convenient orthonormal bases for $H(R)$ and $\mathcal{H}^1(X)$.

Thus, let $T = [0,1]^k$, and X a centered Gaussian process on T with continuous covariance function $R(s,t)$. Let $\lambda_1, \lambda_2, \ldots,$ and $\psi_1, \psi_2, \ldots,$ be, respectively, the eigenvalues and normalised eigenfunctions of the operator $\mathcal{R}\colon \mathcal{L}^2(T) \to \mathcal{L}^2(T)$ defined by $\mathcal{R}\psi(t) = \int_T R(s,t)\psi(s)\,ds$. That is, the λ_n and ψ_n solve the integral equation

$$(3.23)\qquad \int_T R(s,t)\psi(s)\,ds = \lambda\psi(t), \qquad \text{for all } t \in T,$$

and

$$\int_T \psi_n(t)\psi_m(t)\,dt = \begin{cases} 1 & \text{for } n = m, \\ 0 & \text{for } n \neq m. \end{cases}$$

Since it involves no loss of generality, we shall assume that $\lambda_1 \geq \lambda_2 \geq \ldots$.

The following result can be found, for example, in Riesz and Sz-Nagy (1955) when $k = 1$ or Zaanen (1956) for general k.

3.15 THEOREM (MERCER). *Let R, $\{\lambda_n\}_{n\geq1}$ and $\{\psi_n\}_{n\geq1}$ be as above. Then*

$$(3.24)\qquad R(s,t) = \sum_{n=1}^{\infty} \lambda_n \psi_n(s)\psi_n(t),$$

where the series converges absolutely and uniformly on $[0,1]^k \times [0,1]^k$.

The claims made above are best summarised as

3.16 THEOREM. *Let R, $\{\lambda_n\}_{n\geq1}$ and $\{\psi_n\}_{n\geq1}$ be as defined above. Then $\{\sqrt{\lambda_n}\psi_n\}$ is a complete orthormal system in $H(R)$.*

PROOF: Set $\phi_n = \sqrt{\lambda_n}\psi_n$ and define

$$H = \left\{ f\colon f(t) = \sum_{n=1}^{\infty} a_n\phi_n(t),\ t \in [0,1]^k,\ \sum_{n=1}^{\infty} a_n^2 < \infty \right\}.$$

Give H the inner product

$$(f,g)_H = \sum_{n=1}^{\infty} a_n b_n,$$

where $f = \sum a_n \phi_n$ and $g = \sum b_n \phi_n$.

To check that H has the reproducing kernel property, note that

$$\begin{aligned}
(f(\cdot), R(t,\cdot))_H &= \Big(\sum_{n=1}^{\infty} a_n \phi_n(\cdot), \sum_{n=1}^{\infty} \sqrt{\lambda_n}\psi_n(t)\phi_n(\cdot)\Big) \\
&= \sum_{n=1}^{\infty} \sqrt{\lambda_n}\, a_n \psi_n(t) \\
&= f(t).
\end{aligned}$$

It remains to be checked that H is in fact a Hilbert space, and that $\{\sqrt{\lambda_n}\psi_n\}$ is both complete and orthonormal. But all this is standard, given Mercer's theorem, and so is left to you. ■

Remaining with the basic notation of Mercer's theorem, we thus have that the RKHS, $H(R)$, consists of all square integrable functions f on $[0,1]^k$ for which

$$\sum_{n=1}^{\infty} \frac{1}{\lambda_n} \left| \int_T f(t)\psi_n(t)\,dt \right|^2 < \infty,$$

with inner product

$$(f,g)_H = \sum_{n=1}^{\infty} \frac{1}{\lambda_n} \int_T g(t)\psi_n(t)\,dt \int_T g(t)\psi_n(t)\,dt.$$

The Karhunen-Loève expansion of X is obtained by setting $\phi_n = \lambda_n^{\frac{1}{2}}\psi_n$ in the orthonormal expansion (3.17), so that

$$X_t = \sum_{n=1}^{\infty} \lambda_n^{\frac{1}{2}} \xi_n \psi_n(t), \tag{3.25}$$

where the ξ_n are orthonormal Gaussian.

We shall give one, classic, example – that of standard Brownian motion on $[0,1]$. Unfortunately, not too many examples are known, since the integral equation (3.23) is generally not easy to solve. Nevertheless, the fact that it is always possible to solve (3.23) numerically implies that Karhunen-Loève expansions are of substantial practical importance in a variety of applied settings, most notably communication theory.

For Brownian motion (3.23) becomes

$$\begin{aligned}
\lambda\psi(t) &= \int_0^1 \min(s,t)\psi(s)\,ds \\
&= \int_0^t s\psi(s)\,ds + t\int_t^1 \psi(s)\,ds.
\end{aligned}$$

Differentiating both sides with respect to t gives

$$\lambda\psi'(t) = \int_t^1 \psi(s)\,ds,$$
$$\lambda\psi''(t) = -\psi(t),$$

together with the boundary condition $\psi(0) = 0$.

The solutions of this pair of differential equations are given by

$$\psi_n(t) = \sqrt{2}\sin\left(\tfrac{1}{2}(2n+1)\pi s\right), \qquad \lambda_n = \left(\frac{2}{(2n+1)\pi}\right)^2,$$

as is easily verified by substitution. Thus, the Karhunen-Loève expansion of Brownian motion on $[0,1]$ is given by

$$W_t = \frac{\sqrt{2}}{\pi}\sum_{n=0}^{\infty} \xi_n \left(\frac{2}{2n+1}\right) \sin\left(\tfrac{1}{2}(2n+1)\pi t\right),$$

where (ξ_n) is an orthonormal Gaussian sequence.

4. Exercises.

SECTION 3.1:

1.1 Fernique's (1978) original proof of Theorem 3.2 is of intrinsic interest. His argument for showing that the finiteness of $\|X\|$ implies the existence of exponential moments is roughly as follows. You should fill in the details for yourself.

Let X^1 and X^2 be two independent copies of X. Then, for every pair (a,b) of reals

$$P\{\|X^1\| \le a\}\,P\{\|X^2\| > b\}$$
$$= P\{\|X^1 + X^2\| \le \sqrt{2}a,\ \|X^1 - X^2\| > \sqrt{2}b\},$$

from which it follows that

$$P\{\|X^1\| \le a\}P\{\|X^2\| > b\} \le \left(P\{\|X\| > (b-a)/\sqrt{2}\}\right)^2.$$

Choose $a > 0$ such that $q := P\{\|X\| \le a\} \in (\tfrac{1}{2}, 1)$, set

$$b_0 = a, \qquad b_{n+1} = a + b_n\sqrt{2}, \qquad n \ge 0,$$

as well as $x_0 = (1-q)/q < 1$, and define x_n, $n \ge 1$ by

$$P\{\|X\| > b_n\} = qx_n.$$

It follows by induction that

$$P\{\|X\| > b_n\} \;\le\; q\left(\frac{1-q}{q}\right)^{2^n}.$$

Since the recursive definition of the b_n implies that

$$\begin{aligned} b_n &= a\big(1+\sqrt{2}+\cdots+(\sqrt{2})^n\big) \\ &= a(\sqrt{2}+1)(2^{(n+1)/2}-1), \end{aligned}$$

it follows that

$$Ee^{\alpha\|X\|^2} \;\le\; q\left(e^{\alpha a^2} + \sum_{n=0}^{\infty}\left(\frac{1-q}{q}\right)^{2^n} \exp\left(\alpha(\sqrt{2}+1)^2\,(2^{(n+2)/2}-1)^2 a^2\right)\right),$$

which converges for small enough α.

1.2 An interesting side result, that is any easy consequence of the above calculations, is the following:

If X is centered Gaussian, then there exists a universal constant C such that

$$P\{\|X\| \le \lambda\} \ge \tfrac{1}{2} \qquad \Longrightarrow \qquad E\|X\| \le C\lambda.$$

Prove this.

SECTION 3.2:

2.1 Let X be an Ornstein-Uhlenbeck process on $T = [a,b]$, $-\infty < a, b < \infty$; i.e. the centered, stationary Gaussian process on T with covariance function $R(s,t) = e^{-\alpha|s-t|}$. Show that the RKHS of X is made up of all absolutely continuous functions with inner product

$$\begin{aligned} (f,g)_H &= f(a)g(a) + \frac{1}{2\alpha}\int_T \big(f'(t)+\alpha f(t)\big)\big(g'(t)+\alpha g(t)\big)\,dt \\ &= \tfrac{1}{2}\big(f(a)g(a)+f(b)g(b)\big) + \frac{1}{2\alpha}\int_T \big(f'(t)g'(t)+\alpha^2 f(t)g(t)\big)\,dt. \end{aligned}$$

What happens when $T = \Re$?

2.2 Find the RKHS, along with the appropriate inner product, for the Brownian sheet on $[0,1]^k$.

2.3 Here is a very basic zero-one law for vector valued Gaussian variables.

Let X be a $\Re^k$-valued, centered, Gaussian variable, and E a subspace of $\Re^k$. Let X_1 and X_2 be independent copies of X, and for $\theta \in [0,\pi/2]$ set

$$A(\theta) \;=\; \{X_1\cos\theta + X_2\sin\theta \in E,\; X_1\sin\theta - X_2\cos\theta \notin E\}.$$

(i) Show that $P\{A(\theta)\}$ is independent of θ.
(ii) Show that if $\theta_1 \neq \theta_2$, but $x_1, x_2 \in \Re^k$ are such that both $x_1 \cos\theta_1 + x_2 \sin\theta_1 \in E$ and $x_1 \cos\theta_2 + x_2 \sin\theta_2 \in E$, then both $x_1 \in E$ and $x_2 \in E$, and so $x_1 \sin\theta_1 - x_2 \cos\theta_1 \in E$. Conclude that the events $A(\theta)$ are thus disjoint.
(iii) Show that $P\{X \in E\} = 0$ or 1.
(iv) Let $\|X\| = \sup_{i=1,\ldots,k} |X(i)|$. Show that $P\{\|X\| < \infty\} = 0$ or 1.
(v) Note how easy it is to extend the above to X taking values in any vector space Λ on which $\| \; \|$ is a pseudo-semi-norm. (i.e. a mapping from Λ to $(\bar{\Re}, \mathcal{B}(\bar{\Re}))$ for which $(\| \; \|)^{-1}(\Re)$ is a subspace of Λ on which $\| \; \|$ induces a semi-norm.)

2.4 Apply Corollary 3.14 to show that if X is a Gaussian process on a bounded $T \subset \Re$ then with probability one or zero the paths of X
(i) are bounded on T,
(ii) are continuous on T,
(iii) are free of oscillatory discontinuities on T,
(iv) satisfy a Hölder condition on T.

2.5 Complete the proof of the sufficiency part of Theorem 1.5.

SECTION 3.3:

3.1 Let X_t be the centered Gaussian process on $\Re$ with covariance function $R(s,t) = R(t-s) = \cos 2\pi(t-s)$. Find the Karhunen-Loève expansion of X and show that it has only two terms.

IV. BOUNDEDNESS AND CONTINUITY

1. Majorising Measures.

As usual, T is the parameter space of a centered Gaussian process X, equipped with and totally bounded in the canonical metric d. Let m be a probability measure on the Borel subsets of T, and, since we shall need it very often in the following, let $g\colon (0,1] \to \Re_+$ be the function defined by

$$g(t) = \big(\log(1/t)\big)^{\frac{1}{2}}, \qquad 0 < t \le 1. \tag{4.1}$$

Let $B(t,\epsilon)$ be an ϵ-ball in the d-metric about the point $t \in T$.

DEFINITION. *A probability measure m is called a majorising measure (for (T,d)) if*

$$\sup_{t\in T} \int_0^\infty g\big(m(B(t,\epsilon))\big)\, d\epsilon \;<\; \infty. \tag{4.2}$$

Majorising measures were formally introduced by Fernique (1974) (under the name "mesures majorantes"), although in essence they date back to a real variable lemma of Garsia, Rodemich and Rumsey (1970) and a paper of Preston (1972). Fernique proved (4.3) below, thus establishing that the existence of a majorising measure implied the boundedness of $\|X\| = \sup_{t\in T} X_t$. He continued his study of majorising measures in a number of papers, extending the ideas and coming very close to proving that they were the right tool to provide both necessary *and* sufficient conditions for continuity. For example, in Fernique (1978), he obtained both parts of Theorem 4.1 below under additional structural requirements on the space (T,d). In a pathbreaking paper, Talagrand (1987) completed Fernique's program by proving the second part of the following result in the most general case.

In this result, as with all others in this chapter, we shall assume without further comment that T has strictly positive diameter D. (Otherwise $m(B(t,\epsilon)) = 1$ for all ϵ, so that $g(m(B(t,\epsilon))) \equiv 0$, and the following result is trivial.)

4.1 THEOREM. *If m is any probability measure on (T,d), then*

$$E\|X\| \;\le\; K \sup_t \int_0^\infty g\big(m(B(t,\epsilon))\big)\, d\epsilon, \tag{4.3}$$

for some universal constant $K \in (1,\infty)$. Furthermore, K can be chosen such that if X is bounded with probability one, then there exists a probability measure m on (T,d) satisfying

$$K^{-1} \sup_t \int_0^\infty g\big(m(B(t,\epsilon)\big)\, d\epsilon \;\le\; E\|X\|. \tag{4.4}$$

That is, X is a.s. bounded on T if, and only if, (T, d) admits a majorising measure.

The proof is deferred to the following two sections. Note that, as was the case for the entropy integrals which we met in the Introduction, the upper limit of the integrals in (4.3) and (4.4) is really the d-diameter of T. Also note that while this result centers on the a.s. boundedness of X, we have already seen in the previous chapter that this is also the key to the continuity problem (c.f. Theorem 4.5).

The results of Theorem 4.1 are not intuitively immediate, nor are majorising measures easily understood. One thing is clear. For the integrals in (4.3) and (4.4) to be finite, it is necessary that m puts as much mass as possible on regions where the d-balls are small. Since small d-balls imply high incremental variance, these are the regions where the process has the most irregularities in its sample path, and so these are the regions where one would expect the supremum to be achieved. In other words, it is natural to expect that majorising measures are somehow related to the position of the supremum of X. However, this argument has not yet been made rigorous in general. (An obvious problem, of course, is there is no guarantee that in the general case the position of the supremum is uniquely defined.)

To see an example in which there is a strong relationship between majorising measures and the position of the supremum, take T finite and separated by d. Then there is, with probability one, a unique point $\tau(\omega) \in T$ such that $X_\tau(\omega) = \sup_{t \in T} X_t(\omega)$. Let μ_τ denote the law of τ. The following is true:

4.2 THEOREM. *With T, μ_τ, D as above, there exists a universal K such that*

$$(4.5) \qquad K^{-1} E\|X\| \;\le\; D + \int_T \mu_\tau(dt) \int_0^\infty g\big(\mu_\tau(B(t,\epsilon))\big)\, d\epsilon \;\le\; KE\|X\|.$$

Although we shall not prove this result, (the left-hand inequality can be deduced from results in Fernique (1976), the right-hand is due to Talagrand (1987), but neither are far removed from results that we *shall* prove)) it is worthwhile to try to understand it.

Note firstly that the finiteness of the double integral in (4.5) does not imply that μ_τ is a majorising measure. This would be true, however, if *a priori* we had removed from the finite space T all points of μ_τ-measure zero; i.e. points at which the supremum cannot be achieved. The proof of this is immediate once one thinks of the double integral as a finite double sum.

Another way of saying almost the same thing, after a fashion that at first looks as if it would generalise easily to countable parameter spaces and then perhaps to the most general separable ones, (but does not), is the following:

Choose $\alpha > 0$ and set

$$\widehat{T} = \widehat{T}(\alpha) = \Big\{t \in T: \int_0^\infty g\big(\mu_\tau(B(t,\epsilon))\big)\, d\epsilon \le \alpha K\, E\|X\|\Big\}.$$

By (4.5)

$$\begin{aligned} 1 - \mu_\tau(\widehat{T}) &= \mu_\tau(T \setminus \widehat{T}) \\ &= \int_{T\setminus\widehat{T}} \mu_\tau(dt) \\ &\le \int_{T\setminus\widehat{T}} \Big\{\frac{\int_0^\infty g\big(\mu_\tau(B(t,\epsilon))\big)\, d\epsilon}{\alpha K E\|X\|}\Big\} \mu_\tau(dt) \\ &\le \alpha^{-1}. \end{aligned}$$

Thus $\mu_\tau(\widehat{T}) \ge (1-\alpha^{-1})$, and on $\widehat{T}$ the measure μ_τ behaves like a majorising measure, with the unimportant restriction that it is not supported on $\widehat{T}$. We don't know very much about $\widehat{T}$, but, by taking α large, we see that $\widehat{T}$ contains most of the information about $\|X\|$, since

$$P\{\sup_{t\in\widehat{T}} X_t = \sup_{t\in T} X_t\} = \mu_\tau(\widehat{T}) \ge 1 - \alpha^{-1}.$$

Thus, from a heuristic viewpoint, Theorem 4.2 means that μ_τ is a majorising measure on a subset of T large enough to control $\sup_{t\in T} X_t$. ■

It would be interesting to extend Theorem 4.2 to the general case. David Pollard (1990) has recently shown that the supremum of a continuous Gaussian process is attained at a unique point in T. In terms of the results to follow, which will closely relate continuity to the integrals in (4.3) and (4.4), this seems to suggest that a result like Theorem 4.2 should hold in a setting more general than that of finite T. Nevertheless, this does not seem to be an easy extension to establish. ■

Although Theorem 4.2 covers only the case of finite T, it does have a partial extension to general parameter spaces. Take T general, and $\{T_n\}$ a sequence of finite subsets of T increasing to a countable dense subset. By separability and monotonicity

$$E\|X\|_T = \lim_{n\to\infty} E\|X\|_{T_n}.$$

But by Theorem 4.2 we have that, for each n,

$$K^{-1} E\|X\|_{T_n} \le D + \sup_{\mu\in\mathcal{P}(T)} \int_T \mu(dt) \int_0^\infty g\big(\mu(B(t,\epsilon))\big)\, d\epsilon,$$

where $\mathcal{P}(T)$ denotes the collection of all probability measures on T. The fact that the supremum is taken over all $\mathcal{P}(T)$ rather than $\mathcal{P}(T_n)$ only has the effect of making the upper bound larger.

By absorbing the factor of D into the integral – changing the constant if necessary and checking the argument around (4.30) if this bothers you – we have proven the upper bound in

4.3 THEOREM. *There exists a universal constant $K \in (1, \infty)$ such that for any centered Gaussian process X on (T, d)*

$$
\begin{aligned}
K^{-1} \sup_{\mu \in \mathcal{P}(T)} \int_T \mu(dt) \int_0^\infty & g(\mu(B(t,\epsilon)))\, d\epsilon \\
\leq\ & E\|X\| \\
& \leq K \sup_{\mu \in \mathcal{P}(T)} \int_T \mu(dt) \int_0^\infty g(\mu(B(t,\epsilon)))\, d\epsilon.
\end{aligned}
$$

The proof of the lower bound, which is the more interesting, is deferred until the following section.

There is one case in which the majorising measure for the lower bound (4.4) can be easily identified, so that Theorem 4.1 provides a useful necessary and sufficient condition for sample path boundedness. This is the case in which the process is stationary: i.e. T is an abelian group with an operation which shall denote as $+$ such that for all $s, t, \tau \in T$

$$E\, X(s) X(t) \;=\; E\, X(s+\tau) X(t+\tau).$$

4.4 THEOREM. *If X is stationary and T compact, then (4.3) and (4.4) hold with m taken to be normalised Haar measure on T. If T is a compact subset of an infinite group T', then (4.3) and (4.4) hold for the normalised restriction of Haar measure on T' to T.*

PROOF: The inequality (4.3) is true for any measure, and so for the Haar measures of the theorem. Thus we need only prove (4.4). For simplicity, we shall start with the case in which the full group T is compact.

Suppose there exists a majorising measure, m, satisfying (4.4) and define

$$D_m \;=\; \sup\left\{\eta : m(B(t,\eta)) \;<\; \tfrac{1}{2}, \text{ for all } t \in T\right\}.$$

Clearly

$$\int_0^{D_m} g(m(B(t,\epsilon)))\, d\epsilon \;\leq\; K E\|X\|,$$

for all $t \in T$. Let μ denote normalised Haar measure on T, and let τ be a random variable with distribution μ; i.e. τ is uniform on T. For each $\epsilon > 0$, set $Z(\epsilon) = m(B(\tau, \epsilon))$. Then, for any $t_o \in T$

$$\begin{aligned} EZ(\epsilon) &= \int_T m(B(t,\epsilon))\,\mu(dt) \\ &= \int_T m(t + B(t_o,\epsilon))\,\mu(dt) \\ &= \int_T \mu(t + B(t_o,\epsilon))\,m(dt) \\ &= \mu(B(t_o,\epsilon)), \end{aligned}$$

where the second equality comes from the stationarity of X and the third and fourth from the properties of Haar measures.

Now note that $g(x)$ is convex over $x \in (0, \frac{1}{2})$, so that it is possible to define a function $\widehat{g}$ that agrees with g on $(0, \frac{1}{2})$, is bounded on $(\frac{1}{2}, \infty)$, and convex on all of $\Re_+$. By Jensen's inequality,

$$\widehat{g}(EZ(\epsilon)) \le E\widehat{g}(Z(\epsilon)).$$

That is,

$$\widehat{g}(\mu(B(t_o,\epsilon))) \le \int_T \widehat{g}(m(B(t,\epsilon))\,\mu(dt).$$

With D_μ defined for μ as D_m was for m, set $\Lambda = \min(D_m, D_\mu)$. Then

$$\begin{aligned} \int_0^\Lambda \widehat{g}(\mu(B(t_o,\epsilon))\,d\epsilon &\le \int_0^\Lambda d\epsilon \int_T \widehat{g}(m(B(t,\epsilon))\,\mu(dt) \\ &= \int_T \int_0^\Lambda \mu(dt)\widehat{g}(m(B(t,\epsilon))\,d\epsilon \\ &\le KE\|X\|. \end{aligned}$$

This is the crux of (4.4). The final stage of the proof, for this case, is left to you.

The case of T a compact subset of an infinite group T' is handled similarly. Let μ_T be the restriction of Haar measure on T' to T. For convenience, assume that $\mu_T(T) = 1$, so that we need not worry about the normalisation. Then argue as before, but replacing (4.6) by

$$\begin{aligned} EZ(\epsilon) &= \int_T m(B(t,\epsilon))\,\mu_T(dt) \\ &\le \int_{T'} m(t + B(t_o,\epsilon))\,\mu(dt) \\ &= \int_{T'} \mu(t + B(t_o,\epsilon))\,m(dt) \\ &= \mu(B(t_o,\epsilon)). \end{aligned} \tag{4.6}$$

Now continue through the rest of the proof as for the compact case, making certain that the inequalities always go the right way. (Remember that g, and so $\widehat{g}$, are decreasing functions.) ∎

The time has come to look at some applications of the above results. To start, we shall combine Theorem 4.1 with Theorem 3.3 to obtain a simple, applicable, condition for continuity.

For a probability measure m on (T,d) write

$$\gamma_m(\eta) = \sup_{t\in T} \int_0^\eta g\big(m(B(t,\epsilon))\big)\, d\epsilon.$$

4.5 THEOREM. *An centered Gaussian process on T is a.s. bounded and uniformly continuous if, and only if, there exists a probability measure m on (T,d) such that*

$$\lim_{\eta\to 0} \gamma_m(\eta) = 0. \tag{4.7}$$

REMARK: Of course you remember that implicit in this result, as throughout these notes, is the assumption that T is totally bounded. There is, in fact, a somewhat stronger version of Theorem 4.5, which states that X will be a.s. bounded and uniformly continuous on an arbitrary metric space (T,d) if, *and only if*, T is totally bounded in d and (4.7) holds. (Talagrand (1987), Theorem 4.5.) It is this extended result that permitted the claim, at the beginning of Chapter 1, that we lost no interesting cases by restricting ourselves to totally bounded parameter spaces from the start.

HALF PROOF: We shall prove only the sufficiency of (4.7). The necessity is somewhat harder.

As a first step we need to show that (4.7) implies that X is a.s. bounded. Take a $\eta > 0$ such that $\gamma(\eta) < \infty$. Then, by Theorem 4.1, X is a.s. bounded on each ball of radius η. Since T is totally bounded, it can be covered by finitely many such balls. Thus X is bounded on all of T.

To handle continuity, by Theorem 3.3 we need only show $\lim_{\eta\to 0}\phi_d(\eta) = 0$, where

$$\phi_d(\eta) = E \sup_{d(s,t)<\eta} (X_s - X_t). \tag{4.8}$$

The idea of the proof is to bound $\phi_d(\eta)$ via an integral involving a majorising measure, and to then derive convergence to zero from (4.7).

Let $U = \{(s,t) \in T\times T : d(s,t) \le \eta\}$. Provide $T\times T$ and its subspace U with the distance d' given by

$$d'\big((s,t),(s',t')\big) = \left(E[(X_s - X_t) - (X_{s'} - X_{t'})]^2\right)^{\frac{1}{2}}.$$

Note that $X_s - X_t$ is a centered Gaussian process on $T \times T$ with canonical metric d'. Theorem 4.1 should apply to this two-parameter process.

We start with some observations on the difference process. Firstly, note that for $(s,t) \in T \times T$ we have

$$B\big((s,t),\epsilon\big) \supset B(s,\epsilon/2) \times B(t,\epsilon/2), \tag{4.9}$$

where the first ball is a d'-ball and the last two d-balls.

Next, for $x = (s,t) \in T \times T$, let $\phi(x) \in U$ satisfy

$$d'\big(x,\phi(x)\big) \;=\; d'\big(x,U\big) \;=\; \inf_{y\in U} d'\big(x,y\big).$$

(U is closed, so the infimum is achieved.) Note that if $x \in U$ then we can, and shall, take $\phi(x) = x$. Furthermore, if $y \in U$ and $x \in T \times T$ then

$$\begin{aligned} d'\big(y,\phi(x)\big) &\le d'\big(y,x\big) + d'\big(x,\phi(x)\big) \\ &= d'\big(y,x\big) + d'\big(x,U\big) \\ &\le 2d'\big(x,y\big). \end{aligned}$$

Consequently, for all $y \in U$,

$$\phi\big(B_{d'}(y,\epsilon)\big) \subseteq U \cap B_{d'}\big(y,2\epsilon\big). \tag{4.10}$$

Now define a probability measure μ on U by setting $\mu(A) = (m \otimes m) \circ \phi^{-1}(A)$. In view of (4.9) and (4.10) we have, for all $(s,t) \in U$,

$$\begin{aligned} m\big(B(s,\tfrac{1}{2}\epsilon)\big)\, m\big(B(t,\tfrac{1}{2}\epsilon)\big) &= (m\otimes m)\,\big(B(s,\tfrac{1}{2}\epsilon) \times B(t,\tfrac{1}{2}\epsilon)\big) \\ &\le (m\otimes m)\,\big(B_{d'}((s,t),\epsilon)\big) \\ &\le (m\otimes m)\big(\phi^{-1}\big(U \cap (B_{d'}((s,t),2\epsilon)\big)\big) \\ &= \mu\,\big(U \cap (B_{d'}((s,t),2\epsilon)\big). \end{aligned}$$

It thus follows from Theorem 4.1 that for any measure m on T and any $\eta > 0$

$$\begin{aligned} \phi_d(\eta) &\le K \sup_{(s,t)\in U} \int_0^{2\eta} g\big(\mu\big(B_{d'}\big((s,t),\epsilon\big)\big)\big)\, d\epsilon \\ &\le K \sup_{(s,t)\in T\times T} \int_0^{2\eta} g\big(m\big(B_{d'}\big(s,\epsilon/4\big)\big)\, m\big(B_{d'}\big(t,\epsilon/4\big)\big)\big)\, d\epsilon \\ &\le 8K \sup_{t\in T} \int_0^{\eta/2} g\big(m\big(B(t,\epsilon)\big)\big)\, d\epsilon \\ &= 8K\gamma_m\big(\eta/2\big). \end{aligned}$$

This proves the theorem. ■

For necessity, and some further details related to this result, see Talagrand (1987).

As one might expect, majorising measures also feature in providing moduli of continuity, and the following result is the improvement on Corollary 3.4 promised in the previous chapter.

4.6 THEOREM. *Let X be a.s. bounded on T and τ be a metric on T such that the canonical metric d is τ-uniformly continuous. Let $\phi_\tau(\eta)$ be as at (4.8) and denote the τ-modulus of (uniform) continuity of X by*

$$W_\tau(\eta) = \sup_{\tau(s,t)<\eta} |X_s - X_t|.$$

If $\lim_{\eta\to 0}\phi_\tau(\eta) = 0$ then there exists an a.s. finite random variable $\delta = \delta(\omega)$ such that, for almost all ω,

$$W_\tau(\eta) \le \phi_\tau(\eta)$$

for all $\eta \le \delta(\omega)$. That is, $\phi_\tau(\cdot)$ is a uniform sample modulus for X in the metric τ.

This is, of course, Corollary 3.4 without the ϵ. An immediate corollary, in terms of the canonical metric d, is

4.7 COROLLARY. *Let m be a majorising measure for X, a.s. bounded on T. Then there exists a universal constant $K < \infty$ and an a.s. finite random variable $\delta(\omega)$ such that, for almost all ω,*

$$\sup_{s,t:\ d(s,t)<\eta} |X_t - X_s| \le K\gamma_m(\eta).$$

for all $\eta \le \delta(\omega)$.

PROOF OF THEOREM 4.6: Note firstly that if T is finite (in the sense that the number of pairs $s,t \in T$ for which $d(s,t) \neq 0$ is finite) the result is trivially true, by taking $\delta(\omega) = (1-\epsilon)\min\{\tau(s,t) : d(s,t) > 0\}$ for any $\epsilon > 0$. Thus assume that T is infinite.

As noted after the statement of Corollary 3.4, all we need to show is that

$$\lim_{\eta\to 0} \frac{\phi_\tau(\eta)}{d_\tau(\eta)} = \infty, \tag{4.11}$$

where $d_\tau(\eta) = \sup_{\tau(s,t)\le\eta} d(s,t)$.

By (4.4) and the argument applied in the proof of Theorem 4.5 there exists a probability measure on T such that a lower bound to the ratio in (4.11) is given by

$$\begin{aligned}
&\frac{K \sup_{t\in T}\int_0^{d_\tau(\eta)} g\big(m(B(t,\epsilon))\big)\,d\epsilon}{d_\tau(\eta)} \\
&\ge K \sup_{t\in T} g\big(m(B(t,d_\tau(\eta)))\big) \\
&= K \sup_{t\in T} \sqrt{\log\left(\frac{1}{m(B(t,d_\tau(\eta)))}\right)}.
\end{aligned}$$

Since d is τ-continuous, $d_\tau(\eta) \to 0$ as $\eta \to 0$, and so to complete the proof it suffices to show that there exists a $t \in T$ for which

$$m(B(t,\eta)) \to 0 \qquad \text{as } \eta \to 0. \tag{4.12}$$

Suppose that this is not true. That is, there exists a $\delta > 0$ such that for all $t \in T$ $\lim_{\eta \to 0} m(B(t,\eta)) \geq \delta$. It then follows that $m(\{t\}) \geq \delta$ for all $t \in T$. Since m is a probability measure this can only happen if T has at most δ^{-1} points. But T was assumed infinite. Thus (4.12) does hold, and the proof is complete. ∎

We now turn to the proof of Theorem 4.1. This is given in two parts, the upper bound (4.3) first, and then the lower bound (4.4). The upper bound proof is (comparatively) easy, the lower bound proof not so. I heartily recommend skipping both if this is your first reading of this chapter, and coming back to them only after you have read Section 4.4. It is only my childhood training as a mathematician and consequent feelings of guilt at skipping proofs that forces me to include them now and not later.

2. Upper Bound Proof.

The proof of (4.3) that we shall give is closely modelled on Fernique (1978), with some further input from Anderson *et. al.* (1988).

We start with an elementary observation, that we shall use routinely in what follows. The proof is left to you.

4.8 OBSERVATION. *If $f(t)$ is a positive decreasing function then*

$$\sum_{n=1}^{\infty} 2^{-n-1} f(2^{-n}) \leq \int_0^1 f(\epsilon)\, d\epsilon \leq \sum_{n=1}^{\infty} 2^{-n} f(2^{-n}). \tag{4.13}$$

We shall assume throughout this and the following section that diam(V) ≤ 1 so as to simplify some of the notational aspects of the proofs. To see that no loss of generality is involved through this assumption, let $D = \text{diam}(V)$ and set $Y(t) = X(t)/D$. Note that balls in the canonical metrics of X and Y are then related by the fact that

$$B_X(t,\epsilon) = B_Y(t,\epsilon/D).$$

Suppose we succeed in proving (4.3) for the process Y. Then,

$$\begin{aligned}\frac{E\|X\|}{D} &= E\|Y\| \\ &\leq K \sup_t \int_0^\infty g\big(m\big(B_Y(t,\epsilon)\big)\big)\, d\epsilon \\ &= K \sup_t \int_0^\infty g\big(m\big(B_X(t,D\epsilon)\big)\big)\, d\epsilon \\ &= \frac{K}{D} \sup_t \int_0^\infty g\big(m\big(B_X(t,\epsilon)\big)\big)\, d\epsilon.\end{aligned}$$

Thus (4.3) holds also for $\|X\|$, with the same constant K.

We start with an arbitrary probability measure μ and some geometry. Take $n \geq 1$ and let $\{t_{ni}\}_{i=1}^{r_n} \subset T$ be a finite set of distinct points of T such that

$$\sup_{t\in T} \inf_i d(t, t_{ni}) \leq 2^{-n-3}.$$

By reordering, if necessary, we can assume that

$$\mu\big(B(t_{ni}, 2^{-n-2})\big) \geq \mu\big(B(t_{n(i+1)}, 2^{-n-2})\big), \qquad \text{for all } i \geq 1. \tag{4.14}$$

Define subsets $\{C_{ni}\}_{i=1}^{r_n}$ inductively as

$$C_{n1} = B\big(t_{n1}, 2^{-n-2}\big) \tag{4.15}$$

$$C_{ni} = \begin{cases} \emptyset, & \text{if } B(t_{ni}, 2^{-n-2}) \cap \bigcup_{j=1}^{i-1} C_{nj} \neq \emptyset, \\ B(t_{ni}, 2^{-n-2}), & \text{otherwise.} \end{cases} \tag{4.16}$$

Next, define a mapping

$$\pi_n : T \to \{t_{ni}\}_{i=1}^{r_n} \tag{4.17}$$

as follows:

Set $\pi_n(t)$ to be the first t_{ni} for which $d(t, t_{ni}) \leq 2^{-n-3}$ and $C_{ni} \neq \emptyset$. If there is no such i such that the second of these conditions is satisfied, then choose the first t_{ni} such that $d(t, t_{ni}) \leq 2^{-n-3}$, and note that since $C_{ni} = \emptyset$ there exists a maximal $j < i$ such that $C_{nj} \cap B(t_{ni}, 2^{-n-2}) \neq \emptyset$. Define $\pi_n(t)$ to be the corresponding t_{ni}. We denote the n-th such collection of $\{t_{ni}\}$ by $\mathcal{T}_n$, and set $\mathcal{T} = \bigcup_n \mathcal{T}_n$.

The first basic property of this construction that we shall use is that:

$$d(t,\pi_n(t)) \le d(t,t_{ni}) + d(t_{ni},t_{nj}) \le 2^{-n-3} + 2\cdot 2^{-n-2} < 2^{-n}. \tag{4.18}$$

Now note that since for each $t \in T$ and $n \ge 1$ there is a unique $C_{ni} := C_n(t)$, such that $\pi_n(t) = t_{ni} \in C_n(t)$ we can write, with some abuse of notation,

$$\mu_n(t) = \mu(C_n(t)).$$

This leads us to the second property of the construction: If t_{ni} is such that $d(t,t_{ni}) \le 2^{-n-3}$ and $C_{ni} \ne \emptyset$, then

$$\begin{aligned}\mu_n(t) &= \mu(B(t_{ni},2^{-n-2})) && \text{(by (4.15)–(4.16))}\\ &\ge \mu(B(t,2^{-n-3})).\end{aligned}$$

On the other hand, if t_{ni} is such that $d(t,t_{ni}) \le 2^{-n-3}$ and $C_{ni} = \emptyset$, then there is a $j < i$ such that $C_{nj} \cap B(t_{ni},2^{-n-2}) \ne \emptyset$, and in this case

$$\begin{aligned}\mu(B(t,2^{-n-3})) &\le \mu(B(t_{ni},2^{-n-2}))\\ &\le \mu(B(t_{nj},2^{-n-2})) && \text{(by (4.14))}\\ &= \mu_n(t).\end{aligned}$$

The above two inequalities yield that in all cases

$$\mu_n(t) \ge \mu(B(t,2^{-n-3})). \tag{4.19}$$

This is all the geometry that we shall require for the moment. We now build ourselves a new, somewhat simpler process than the original. We start with $\{\xi_t\}_{t\in\mathcal{T}}$, a collection of i.i.d. standard Gaussian variables on $\mathcal{T}$. (If a $t \in \mathcal{T}$ appears in more than one $\mathcal{T}_n$ choose a different ξ_t for each n, and change the following proof, where necessary, to account for this irritating phenomenon. We shall assume, however, that this never happens.) Define a new Gaussian process Y on T by

$$Y(t) = \sum_{n=1}^{\infty} 2^{-n}\xi(\pi_n(t)). \tag{4.20}$$

Note that although Y generates its own canonical metric on T, we shall treat it as defined on T equipped with the canonical metric of X. The following step in the proof is important enough to be singled out as an independent result.

4.9 LEMMA. *If Y is defined as at (4.20) then, for all $s,t \in T$,*

$$E(X_s - X_t)^2 \leq 6E(Y_s - Y_t)^2.$$

PROOF: For $s,t \in T$, choose $N \geq 1$ such that

$$2^{-N} < d(s,t) = \left(E(X_t - X_s)^2\right)^{\frac{1}{2}} \leq 2^{-N+1}. \tag{4.21}$$

By (4.18), it thus follows that for $n \geq N+1$, $\pi_n(t) \neq \pi_n(s)$. (For $n < N+1$ it is impossible to say whether $\pi_n(t) = \pi_n(s)$ or not. This is what makes it impossible to use this argument to obtain a proof of the lower bound as well.) Using the representation (4.20) of Y we thus deduce that

$$\begin{aligned} E(Y_t - Y_s)^2 &\geq \sum_{n=N+1}^{\infty} 2^{-2n} E\left[\xi(\pi_n(t)) - \xi(\pi_n(s))\right]^2 \\ &= 2 \sum_{n=N+1}^{\infty} 2^{-2n} \\ &= 2^{-2N+2}/6. \end{aligned}$$

Comparing this to the upper bound in (4.21) proves the lemma. ∎

It now follows from Theorem 2.8 or 2.9 (the two forms of the Sudakov-Fernique inequality) that $E\|X\| \leq 2\sqrt{6}\, E\|Y\|$, and so it suffices to prove (4.3) for the process Y.

To do this, let $\tau\colon \Omega \to T$ be a random point in T. If we can show that there exists an $M > 0$ such that for *every* such random variable τ

$$EY_\tau \leq M, \tag{4.22}$$

then it must follow (by contradiction) that

$$E\|Y\| = E \sup_{t\in T} Y_t \leq M. \tag{4.23}$$

Thus, consider (4.22) and let the distribution of τ on T be given by a probability measure ν. Since each $\mathcal{T}_n$ is finite, (4.20) gives us that

$$EY_\tau = \sum_{n=1}^{\infty} 2^{-n} \sum_{t\in \mathcal{T}_n} E\{\xi_t I_{\{\pi_n(\tau)=t\}}\}. \tag{4.24}$$

Note the easy fact that for ξ standard normal

$$\begin{aligned} E\{\xi I_{\{\xi > \sqrt{2g(a)}\}}\} &= (2\pi)^{-\frac{1}{2}} \int_{\sqrt{2g(a)}}^{\infty} x e^{-\frac{1}{2}x^2}\, dx \\ &= (2\pi)^{-\frac{1}{2}} \left[-e^{-x^2/2}\right]_{\sqrt{2\log(1/a)}}^{\infty} \\ &= (2\pi)^{-\frac{1}{2}} a. \end{aligned} \tag{4.25}$$

Now note that each expectation in the summation in (4.24) can be written as

$$\begin{aligned} &E\{\xi_t I_{\{\pi_n(\tau)=t\}} I_{\{\xi_t>\sqrt{2}g(\mu_n(t))\}}\} + E\{\xi_t I_{\{\pi_n(\tau)=t\}} I_{\{\xi_t\le\sqrt{2}g(\mu_n(t))\}}\} \\ (4.26)\qquad &\le E\{\xi_t I_{\{\xi_t>\sqrt{2}g(\mu_n(t))\}}\} + \sqrt{2}g(\mu_n(t))E\{I_{\{\pi_n(\tau)=t\}}\} \\ &= (2\pi)^{-\frac{1}{2}}\mu_n(t) + \sqrt{2}g(\mu_n(t))\,P\{\pi_n(\tau)=t\}, \end{aligned}$$

the last line following from (4.25) and the definition of μ.

Thus, substituting into (4.24), and applying (4.19) to obtain the third inequality below, we obtain

$$\begin{aligned} EY_\tau &\le \frac{1}{\sqrt{2\pi}} + \sum_{n=1}^{\infty} 2^{-n+\frac{1}{2}} \sum_{t\in T_n} g(\mu_n(t))\,P\{\pi_n(\tau)=t\} \\ (4.27)\qquad &\le \frac{1}{\sqrt{2\pi}} + \sum_{n=1}^{\infty} 2^{-n+\frac{1}{2}} \int_T g(\mu(C_n(t)))\,\nu(dt) \\ &\le \frac{1}{\sqrt{2\pi}} + \sum_{n=1}^{\infty} 2^{-n+\frac{1}{2}} \int_T g(\mu(B(t,2^{-n-3})))\,\nu(dt) \\ &\le \frac{1}{\sqrt{2\pi}} + \int_T \nu(dt) \sum_{n=1}^{\infty} 2^{-n+\frac{1}{2}} g(\mu(B(t,2^{-n-3}))). \end{aligned}$$

Replacing the sum over n by an integral over ϵ (c.f. Observation 4.8) gives us that there exists a constant K such that

$$K^{-1}EY_\tau \le 1 + \int_T \nu(dt) \int_0^\infty g(\mu(B(t,\epsilon)))\,d\epsilon,$$

and so, allowing K to change from line to line,

$$\begin{aligned} E\|X\| &\le K \sup_\tau EY_\tau \\ &= K\Big(1 + \sup_\nu \int_T \nu(dt) \int_0^\infty g(\mu(B(t,\epsilon)))\,d\epsilon\Big) \\ &\le K\Big(1 + \sup_t \int_0^\infty g(\mu(B(t,\epsilon)))\,d\epsilon\Big), \end{aligned}$$

the last inequality coming from the fact that ν is a probability measure.

I claim that $\sup_t \int_0^\infty g(\mu(B(t,\epsilon)))\,d\epsilon$ can be bounded from below by $\sqrt{\log 2}$. To see this, note that T has at least two distinct points (since we have assumed that $\operatorname{diam}(T) > 0$). Thus there must be at least one point $t \in T$ with $\mu(B(t,\epsilon)) \le \frac{1}{2}$ for all $\epsilon < 1$. For this t

$$\begin{aligned}(4.28)\qquad \int_0^\infty g\big(\mu\big(B(t,\epsilon)\big)\big)\,d\epsilon \;&\ge\; \int_0^1 \sqrt{\log 2}\,d\epsilon \\ &=\; \sqrt{\log 2},\end{aligned}$$

as claimed.

Thus the 1 in the above bound can be absorbed into the integral, albeit with a slightly larger constant, and we are done. ■

Although this section is advertised as being devoted to upper bounds, while we have the appropriate notation at hand we shall also give the

PROOF OF THE LOWER BOUND OF THEOREM 4.3: We shall need two facts. The first, as you can verify with a little calculus, is that the function g satisfies

$$(4.29)\qquad x\,g(x) \;\le\; x\,g(y) + \frac{y}{\sqrt{2e}},$$

for all x and y. The second is the rather useful inequality that

$$(4.30)\qquad \sup_{\nu\in\mathcal{P}(T)} \int_T \nu(dt) \int_0^\infty g\big(\nu\big(B(t,\epsilon)\big)\big)\,d\epsilon \;\ge\; \sqrt{\log 2},$$

which you can verify by choosing ν to be concentrated on two points $s, t \in T$ with $d(s,t) > 0$ and $\nu(\{s\}) = \nu(\{t\}) = \frac{1}{2}$. (This is the second, and last, time we need the assumption that T has non-zero diameter.)

In what follows, given a measure $\nu \in \mathcal{P}(T)$ we write τ_ν to denote a T-valued random variable with distribution ν. We shall use the notation of the previous proof without comment other than to point out now that when notation from there appears with a suffix m this is to indicate that the construction of the proof relates to the specific measure m. By (4.4), a proof of which is in the following section, there exists a universal K and a measure m such that

$$\begin{aligned}K\,E\|X\| \;&\ge\; \sup_{t\in T}\int_0^\infty g\big(m\big(B(t,\epsilon)\big)\big)\,d\epsilon \\ &=\; \sup_{\nu\in\mathcal{P}(T)} \int_T \nu(dt)\int_0^\infty g\big(m\big(B(t,\epsilon)\big)\big)\,d\epsilon \;-\; \eta,\end{aligned}$$

for any $\eta > 0$. To prove the above inequality simply take ν concentrated at a point t_o for which

$$\int_0^\infty g\big(m\big(B(t_o,\epsilon)\big)\big)\,d\epsilon \;\ge\; \sup_{t\in T}\int_0^\infty g\big(m\big(B(t,\epsilon)\big)\big)\,d\epsilon \;-\; \eta.$$

Thus

$$
\begin{aligned}
K\,E\|X\| + \eta &\geq \sup_{\nu\in\mathcal{P}(T)} \int_T \nu(dt) \sum_{n=1}^{\infty} 2^{-n-4} g\big(m\big(B(t, 2^{-n-3})\big)\big) \\
&\geq \sup_{\nu\in\mathcal{P}(T)} \int_T \nu(dt) \sum_{n=1}^{\infty} 2^{-n-4} g\big(m\big(C_n^m(t)\big)\big) \\
&= \sup_{\nu\in\mathcal{P}(T)} \sum_{n=1}^{\infty} \sum_{t\in\mathcal{T}_n} 2^{-n-4} g\big(m\big(C_n^m(t)\big)\big)\, P\{\pi_n^m(\tau_\nu) = t\}.
\end{aligned}
$$

By (4.29) this is bounded below by

$$
\sup_{\nu\in\mathcal{P}(T)} \sum_{n=1}^{\infty} \sum_{t\in\mathcal{T}_n} 2^{-n-4} \Big[g\big(P\{\pi_n^m(\tau_\nu) = t\}\big)\, P\{\pi_n^m(\tau_\nu) = t\} - \frac{m\big(C_n^m(t)\big)}{\sqrt{2e}} \Big]
$$

which, by (4.18), is in turn bounded below by

$$
\sup_{\nu\in\mathcal{P}(T)} \sum_{n=1}^{\infty} \sum_{t\in\mathcal{T}_n} 2^{-n-4} g\big(\nu\big(B(t, 2^{-n})\big)\big)\, P\{\pi_n^m(\tau_\nu) = t\} - (16\sqrt{2e})^{-1}.
$$

The sum is equal to

$$
\begin{aligned}
&\sup_{\nu\in\mathcal{P}(T)} \sum_{n=1}^{\infty} 2^{-n-4} E\Big\{ g\big(\nu\big(B(\pi_n^m(\tau_\nu), 2^{-n})\big)\big) \Big\} \\
&\geq \sup_{\nu\in\mathcal{P}(T)} \sum_{n=1}^{\infty} 2^{-n-4} E\Big\{ g\big(\nu\big(B(\tau_\nu, 2^{-n+1})\big)\big) \Big\} \\
&\geq \sup_{\nu\in\mathcal{P}(T)} \frac{1}{16} \int_0^{\infty} E\Big\{ g\big(\nu\big(B(\tau_\nu, 2\epsilon)\big)\big) \Big\}\, d\epsilon \\
&\geq \sup_{\nu\in\mathcal{P}(T)} \frac{1}{32} \int_T \nu(dt) \int_0^{\infty} g\big(\nu\big(B(t,\epsilon)\big)\big)\, d\epsilon.
\end{aligned}
$$

Now apply (4.30) to find that, modulo a multiplicative constant, $E\|X\|$ is bounded below by

$$
\sup_{\nu\in\mathcal{P}(T)} \int_T \nu(dt) \int_0^{\infty} g\big(\nu\big(B(t,\epsilon)\big)\big)\, d\epsilon,
$$

which is precisely what we had to prove. ∎

3. Lower Bound Proof.

The time has now come to tackle the hardest proof in these notes, that of establishing the lower bound in Theorem 4.1.

In principle, the idea behind the proof is easy. Recall that the proof of the upper bound was based on comparing a given process X with a simpler process Y which dominated it, in the sense that $E(X_t - X_s)^2 \le \text{const}\, E(Y_t - Y_s)^2$ for all $s, t \in T$. Since the structure of Y was simpler, it was possible to calculate a good upper bound for $E\|Y\|$, and so, by the Sudakov-Fernique inequality, obtain an upper bound for $E\|X\|$ as well.

We shall follow this path once again. What makes it somewhat harder to traverse this time, however, is that even after we have found a simple comparison process, the lower bound calculation is difficult. *C'est la vie.* What is somewhat more disappointing is that even after we shall have worked so hard to prove the result, we shall not have discovered how to built majorising measures. The proof is not constructive.

Before we start, we require the following standard result.

4.10 LEMMA. *Let $X_1, \ldots, X_N$ be a sequence of i.i.d. standard Gaussian variables. There exists a universal $K > 0$ such that*

$$E \sup_{1 \le i \le N} X_i \ge K\sqrt{\ln N}. \tag{4.31}$$

PROOF: Take $Z_1, \ldots, Z_N$ to be a sequence of i.i.d. random variables on $[1, \infty]$ with probability density $f(z) = z^{-2}$ and distribution function $F(z) = 1 - z^{-1}$.

The probability density of the $\min(Z_1, \ldots, Z_N)$ is thus given by

$$Nf(z)(1 - F(z))^{N-1} = Nz^{-N-1},$$

so that

$$E \min(Z_1, \ldots, Z_N) = \frac{N}{N-1}.$$

Note that $Z_i \stackrel{\mathcal{L}}{=} 1/\Psi(X_i)$, where Ψ, as usual, denotes the standard normal distribution function. Thus

$$\begin{aligned} \frac{N}{N-1} &= E \min(Z_1, \ldots, Z_N) \\ &= E \min\left\{\frac{1}{\Psi(X_1)}, \ldots, \frac{1}{\Psi(X_N)}\right\} \\ &= E\left(\Psi(\max(X_1, \ldots, X_N))\right)^{-1}. \end{aligned} \tag{4.32}$$

But $\Psi^{-1}(x)$ is a convex function, since

$$\begin{aligned}\frac{d^2\Psi^{-1}(x)}{dx^2} &= \frac{d}{dx}\Big(-\phi(x)\Psi^{-2}(x)\Big)\\ &= \phi(x)\Psi^{-3}(x)\,\big(2\phi(x)+x\Psi(x)\big),\end{aligned}$$

and the last expression is clearly positive for $x>0$ and also positive for $x<0$ since it is zero at $-\infty$ and increasing thereafter.

Thus Jensen's inequality and (4.32) give

$$\Psi\Big(E\max_{1\le i\le N} X_i\Big) \ge \frac{N-1}{N}.$$

Apply the inequalities of (2.1) to complete the proof. ■

We now require some rather formidable notation, in which A will always be a subset of T, m a measure on T and $\mathcal{P}(A)$ the set of probability measures on A.

$$\begin{aligned}\gamma_m(A) &= \gamma(m,A) = \sup_{t\in A}\int_0^\infty g\big(m(B(t,\epsilon))\big)\,d\epsilon\\ \gamma(A) &= \inf_{m\in\mathcal{P}(A)}\gamma_m(A)\\ \alpha(A) &= \sup_{B\subseteq A}\gamma(B)\\ \alpha &= \sup\big\{\alpha(V)\colon \emptyset\neq V\subseteq T,\ V \text{ finite}\big\}.\end{aligned}$$

We have to establish the existence of a $m\in\mathcal{P}(T)$ such that for some universal constant K

$$\gamma_m(T) \le K\,E\|X\|_T.$$

Our first step towards this will be a little more modest.

4.11 LEMMA. *There exists a universal constant K such that for every non-empty, finite, $V\subseteq T$*

$$\alpha(V) \le K\,E\|X\|_V. \tag{4.33}$$

You should think of this lemma as an approximate version of our theorem. It is going to be easier to prove than the general result because of the fact that $\alpha(V)$ is determined by what X does on a finite subset of T. Before we prove this lemma, however, we shall see that it is in fact equivalent to the full result.

PROOF OF THE LOWER BOUND (4.3): Since (4.33) implies that $\alpha\le KE\|X\|_T$, we need only show the existence of a $m\in\mathcal{P}(T)$ for which $\gamma_m(T)\le$

$K\alpha$. This we shall do by a reasonably straightforward subsequencing argument; i.e. We shall use the definitions of the α's and γ's on finite subsets, where infima and suprema are achieved, to show that appropriate m's can be constructed there, and then take a weak limit of these as our subspaces grow towards the whole of T.

For each $n \geq 1$ let $T_n = \{t_{ki}\}_{i=1}^{r_n}$ be a finite subset of T such that

$$\sup_{t \in T} \inf_{1 \leq i \leq r_n} d(t, t_{ki}) \leq 2^{-n-4}.$$

Now we choose one of these sets, which we denote by T_k. (Be careful in following the subscripts in what follows. T_n and T_k are chosen from one large family of sets, but whereas T_k is (temporarily) fixed, T_n will vary.)

Since $\gamma(T_k) \leq \alpha(T_k) \leq \alpha$, there exists a $m_k \in \mathcal{P}(T_k)$ such that, by Observation 4.8,

$$\begin{aligned} \sup_{t \in T_k} \sum_{i=1}^{\infty} 2^{-i} g\big(m_k\big(B(t, 2^{-i})\big)\big) &\leq 2 \sup_{t \in T_k} \int_0^1 g\big(m_k\big(B(t, \epsilon)\big)\big)\, d\epsilon \\ &\leq 2\alpha. \end{aligned} \tag{4.34}$$

Consider m_k as a measure on T, by setting $m_k(A) = m_k(A \cap T_k)$ for $A \subset T$. Now follow the same construction as used in the proof of the upper bound – viz. (4.14)-(4.17) – to obtain a family of subsets $\{C_{ni}^k\}_{i=1}^{r(k,n)}$ of T and a mapping $\pi_n^k : T \to T_k$ satisfying (4.18) and such that for each $t \in T$ the unique $C_n^k(t) = C_{ni}^k$ with center $\pi_n^k(t) = t_{ni}^k$ satisfies

$$m_k\big(C_n^k(t)\big) \leq m_k\big(B(t, 2^{-n-3})\big). \tag{4.35}$$

(As before, we must assume that the $\{t_{ni}^k\}$ have been reordered to satisfy (4.14) for $\mu = m_k$, but, as before, this involves no loss of generality.)

Define $\mu_n^k \in \mathcal{P}(T_n)$ by

$$\mu_n^k(t_{ni}^k) = \frac{m_k(C_{ni}^k)}{\sum_{i=1}^{r_k} m_k(C_{ni}^k)}.$$

Now switch the sequencing, and hold n fixed while sending $k \to \infty$. Since n is fixed, T_n is finite, and so a standard argument shows that there is a subsequence $\{\mu_n^{k'}\}_{k'}$ of the above measures for which $\lim_{k' \to \infty} \mu_n^{k'} = \mu_n$ exists (pointwise) and is an element of $\mathcal{P}(T_n)$.

We are almost done. Fix $t \in T$ and choose a further subsequence such that $\lim_{k'' \to \infty} \pi_n^{k''}(t) = \pi_n(t)$ exists. Note from the finiteness of T_n that for large enough k'' we must have that $\pi_n^{k''}(t) \equiv \pi_n(t)$. Let $\tau_k(t)$ be the first t_{ki} such that $d(t, t_{ki}) \le 2^{-k-4}$. Then, by (4.33), (4.34) and Fatou's lemma

$$\begin{aligned}
\sum_{i=1}^{\infty} 2^{-n} g\big(\mu_n(\pi_n(t))\big) &\le \sum_{i=1}^{\infty} 2^{-n} \liminf_{k''\to\infty} g\Big(\frac{m_{k''}(C_{ni}^{k''}(t))}{\sum_{i=1}^{r_k} m_k(C_{ni}^k)}\Big) \\
&\le \sum_{i=1}^{\infty} 2^{-n} \liminf_{k''\to\infty} g\big(m_{k''}(C_n^{k''}(t))\big) \\
&\le 16 \sum_{i=1}^{\infty} 2^{-n-4} \liminf_{k''\to\infty} g\big(m_{k''}(B(\tau_{k''}(t), 2^{-n-4}))\big) \\
&\le 32\,\alpha.
\end{aligned}$$

Finally, define $m \in \mathcal{P}(T)$ by $m(A) = \sum_{i=1}^{\infty} 2^{-n} \mu_n(A)$, Note that

$$\begin{aligned}
g(m(\{\pi_n(t)\})) &\le g(2^{-n} \mu_n(\{\pi_n(t)\})) \\
&\le g(2^{-n}) + g(\mu_n(\{\pi_n(t)\}))
\end{aligned}$$

and that by (4.18) and (4.35) $m(B(t, 2^{-n})) \ge m(\{\pi_n(t)\})$. Thus

$$\begin{aligned}
\int_0^1 g\big(m(B(t,\epsilon))\big)\, d\epsilon &\le \sum_{i=1}^{\infty} 2^{-n} g\big(m(B(t, 2^{-n}))\big) \\
&\le \sum_{i=1}^{\infty} 2^{-n} g\big(m(B(t, 2^{-n}))\big) \\
&\le \sum_{i=1}^{\infty} 2^{-n} g\big(\mu_n(\{\pi_n(t)\})\big) + \sum_{i=1}^{\infty} 2^{-n} g(2^{-n}) \\
&\le 32\,\alpha + 2\sqrt{\log 2} \\
&\le K\alpha.
\end{aligned}$$

Since t was arbitrary, the same is true for the supremum over t. This completes the proof of the majorising measure lower bound, assuming Lemma 4.11. ∎

We must now show that Lemma 4.11 is valid. Before we can do so, however, we have two preliminary results to establish. These are actually the key to the proof, for they will show how to pack into the metric space (T, d) small balls that are separated from one another by an appropriate (even smaller) distance while at the same time ensuring that many side conditions are satisfied. The proof of Lemma 4.11 then proceeds by replacing X by a process defined, essentially, on the centers of these small balls, and studying the supremum of the new process. The fact that the values of the new process have a certain minimal correlation between them (i.e. the centers of these balls are a certain d-distance apart) is what makes the new process easier to handle.

If you are familiar with mixing arguments for proving central limit or Poisson limit theorems, in which one takes larger and larger time intervals, split into more and more large subintervals, themselves separated by yet more comparatively small but actually quite large subintervals, then you should think of what we are about to do as being a similar construction, but with "small" replacing "large" throughout.

4.12 LEMMA. *If (T,d) is a metric space of diameter D and $A_1,\ldots,A_n$ a partition of T, then there exists a non-empty subset I of $\{1,\ldots,n\}$ such that, for all $i \in I$,*

$$\alpha(A_i) \geq \alpha(T) - D\sqrt{2\log(1+|I|)},$$

where $|I|$ denotes the number of elements in I.

PROOF: Order the A_i so that $\alpha(A_1) \geq \alpha(A_2) \geq \ldots$. Fix $S \subset T$, and choose $m_i \in \mathcal{P}(A_i \cap S)$, $i = 1,\ldots,n$. Set $a_i = (i+1)^{-2}$, note $\sum_{i=1}^n a_i \leq 1$, and define $m \in \mathcal{P}(S)$ by

$$m(A) = \frac{\sum_{i=1}^n a_i m_i(A \cap A_i)}{\sum_{i=1}^n a_i m_i(S)}.$$

Then, for $t \in S \cap A_i$,

$$\begin{aligned}\int_0^\infty g(m(B(t,\epsilon)))\,d\epsilon &\leq \int_0^D g(a_i m_i(B(t,\epsilon)))\,d\epsilon \\ &\leq Dg(a_i) + \int_0^D g(m_i(B(t,\epsilon)))\,d\epsilon.\end{aligned}$$

Thus,

$$\begin{aligned}\gamma(m,S) &\leq \sup_{1\leq i\leq n}\left\{Dg(a_i) + \gamma(m_i, A_i \cap S)\right\} \\ &= Dg(a_{i_o}) + \gamma(m_{i_o}, A_{i_o} \cap S),\end{aligned}$$

for some $i_o \in \{1,\ldots,n\}$. By taking infima first over $m_i \in \mathcal{P}(A_{i_o})$ and then over $m \in \mathcal{P}(S)$ it follows from the above and the definition of γ that $\gamma(S) \leq Dg(a_{i_o}) + \gamma(A_{i_o} \cap S)$. This, in turn, implies that

$$\alpha(T) \leq Dg(a_{i_o}) + \alpha(A_{i_o}).$$

Taking $I = \{1,\ldots,i_o\}$, so that $|I| = i_o$, and noting the monotonicity of the $\alpha(A_i)$, completes the proof. ■

The next lemma is the important one. To formulate it, for each $i \geq 1$ and $A \subseteq T$ set

$$\beta_i(A) \;=\; \alpha(A) \;-\; \sup_{t\in A} \alpha\big(A \cap B(t, 6^{-i-1})\big).$$

4.13 LEMMA. *If (T,d) is a finite metric space of diameter $D \leq 6^{-i}$ then there exists a non-empty subset I of $\{1,\ldots,|T|\}$ and subsets $\{B_k\}_{k\in I}$ of T such that*

$$\operatorname{diam}(B_k) \;\leq\; 6^{-i-1} \tag{4.36}$$

$$d(B_j, B_k) \;\geq\; 6^{-i-2} \qquad \text{for } j \neq k, \tag{4.37}$$

$$\alpha(B_k) + \beta_{i+1}(B_k) \;\geq\; \alpha(T) + \beta_i(T) - 6^{-i+1}\big(2 + \sqrt{\log|I|}\big). \tag{4.38}$$

PROOF: By induction over k, construct a sequence $\{A_k, B_k, T_k\}_k$ of triplets of subsets of T that satisfy

$$T_k = T \setminus \bigcup_{j<k} A_j, \qquad\qquad \emptyset \neq B_k \subseteq A_k \subseteq T_k,$$

$$\operatorname{diam}(A_k) \leq 6^{-i-1}, \qquad\qquad d(B_k, T_k \setminus A_k) \geq 6^{-i-2}$$

and such that for each k either

$$\alpha(B_k) + \alpha(T_k) \;\geq\; 2\,\alpha(A_k), \tag{4.39}$$

or

$$\alpha(B_k) + \beta_{i+1}(B_k) \;\geq\; \alpha(T_k). \tag{4.40}$$

Denote those k for which (4.39) is satisfied by I_1, and those for which (4.40) is satisfied by I_2.

The construction starts with $T_1 = T$. Given T_k, if

$$\alpha\big(B(t, 6^{-i-2})\big) + \alpha(T_k) \;\geq\; 2\,\alpha\big(B(t, 2\cdot 6^{-i-2}) \cap T_k\big) \tag{4.41}$$

for some $t \in T_k$ then set $A_k = B(t, 2\cdot 6^{-i-2}) \cap T_k$ and $B_k = B(t, 6^{-i-2}) \cap T_k$. In this case $k \in I_1$. Otherwise, set $A_k = B(t_o, 3\cdot 6^{-i-2}) \cap T_k$ and $B_k = B(t_o, 2\cdot 6^{-i-2}) \cap T_k$, where $t_o \in T_k$ maximises $\alpha\big(B(t, 2\cdot 6^{-i-2}) \cap T_k\big)$. In this case, $k \in I_2$, since

$$\begin{aligned}
\beta_{i+1}(B_k) \;&\geq\; \alpha(B_k) \;-\; \sup_{t\in B_k} \alpha\big(B(t, 6^{-i-2}) \cap T_k\big)\\
&\geq\; \alpha(B_k) + \alpha(T_k) - 2\sup_{t\in B_k} \alpha\big(B(t, 6^{-i-2}) \cap T_k\big)\\
&\geq\; \alpha(T_k) - \alpha(B_k),
\end{aligned}$$

the second inequality coming from (4.41) with the inequality sign reversed. The construction stops when $\alpha(T_k) < \alpha(T) - 2 \cdot 6^{-i}$, and let κ denote the first k for which this is true. Thus, in particular, $\alpha(T_k) \geq \alpha(T) - 2 \cdot 6^{-i}$ for all $k < \kappa$.

Since $\alpha(T_\kappa) < \alpha(T) - 6^{-i}\sqrt{2\log 3}$, the previous lemma, applied to the two set partition $\{\bigcup_{k<\kappa} A_k, T_\kappa\}$ of T, gives us that

$$\alpha\Big(\bigcup_{k<\kappa} A_k\Big) \geq \alpha(T) - 6^{-i}\sqrt{2\log 2}.$$

Another application of the previous lemma gives that there exists a non-empty $I \subseteq \{1, \ldots, \kappa - 1\} = I_1 \cup I_2$ such that for $k \in I$

$$\begin{aligned}
\alpha(A_k) &\geq \alpha\Big(\bigcup_{k<\kappa} A_k\Big) - 6^{-i}\sqrt{2\log(1+|I|)} \\
&\geq \alpha(T) - 6^{-i}\Big(\sqrt{2\log 2} + \sqrt{2\log(1+|I|)}\Big) \\
&\geq \alpha(T) - 2 \cdot 6^{-i}\Big(2 + \sqrt{\log(1+|I|)}\Big).
\end{aligned}$$

Finally, we obtain that for $k \in I \cap I_1$

$$\begin{aligned}
\alpha(B_k) + \beta_{i+1}(B_k) &\geq 2\,\alpha(A_k) - \alpha(T_k) \\
&\geq \alpha(T) + \beta_i(T) - 3\big(\alpha(T) - \alpha(A_k)\big) \\
&\geq \alpha(T) + \beta_i(T) - 6^{-i}\Big(2 + \sqrt{\log(1+|I|)}\Big).
\end{aligned}$$

Similarly, if $k \in I \cap I_2$ then

$$\begin{aligned}
\alpha(B_k) + \beta_{i+1}(B_k) &\geq \alpha(T) - 2 \cdot 6^{-i} \\
&\geq 2\,\alpha(T) - \alpha(A_k) - 4 \cdot 6^{-i}\Big(2 + \sqrt{\log(1+|I|)}\Big) \\
&\geq \alpha(T) + \beta_i(T) - 6^{-i}\Big(2 + \sqrt{\log(1+|I|)}\Big),
\end{aligned}$$

which proves the lemma. ■

All that now remains is to provide a

PROOF OF LEMMA 4.11: Choose a finite $V \subset T$. Suppose $\alpha(V) \leq 6^3\,\mathrm{diam}(V)$, and take $s, t \in v$ with $d(s,t) = \mathrm{diam}(V)$. Since by Lemma 2.7 $E\max(X_s, X_t) = \mathrm{diam}(V)/\sqrt{2\pi}$, we have that (4.33) holds with $K = \sqrt{2\pi}\,6^3$.

Thus, we assume henceforth that $\alpha(V) > 6^3\,\mathrm{diam}(V)$. Recall that we have a global assumption, as discussed at the beginning of the previous section, that $\mathrm{diam}(V) \leq 1$. Set

$$\begin{aligned}
L &= \max\{j : \mathrm{diam}(V) \leq 6^{-j}\}, \\
M &= \min\Big\{m : \inf_{s,t \in V}\{d(s,t) : d(s,t) > 0\} > 2^{-m}\Big\}.
\end{aligned}$$

We now call on the construction of Lemma 4.13, which can be applied $M-L$ times to establish the existence on non-empty families $\mathcal{B}_L,\ldots,\mathcal{B}_M$ of non-empty subsets of V such that $\mathcal{B}_L=\{V\}$, $\sup_{B\in\mathcal{B}_i}\operatorname{diam}(B)\le 6^{-i}$ and if $B\ne B'$ are both in $\mathcal{B}_i$ then $d(B,B')\ge 6^{-i-2}$. Indeed, a careful sequential application of Lemma 4.13 allows us to assume that for every $B\in\mathcal{B}_i$ and $j<i$ there is a $B'\in\mathcal{B}_j$ with $B\subseteq B'$.

Furthermore, if for $t\in V$ there is a (necessarily unique) $B\in\mathcal{B}_i$ with $t\in B$, then we denote this B by B_t^i and we shall write $N_t^i=|\{B\in\mathcal{B}_{i+1}:B\subseteq B_t^i\}|$. Then, by Lemma 4.13,

$$\alpha(B)+\beta_{i+1}(B)\ \ge\ \alpha(B_t^i)+\beta_i(B_t^i)-6^{-i+1}\left(2+\sqrt{\log N_t^1}\right), \tag{4.42}$$

for each $B\in\mathcal{B}_{i+1}$ such that $B\subset B_t^i$.

Now set $\widetilde{V}=\bigcap_{i=L}^M\bigcup_{B\in\mathcal{B}_i}B$, and $\Psi_t^k=\sum_{i=k}^{M-1}6^{-i-1}\sqrt{\log N_t^i}$, for $t\in\widetilde{V}$. Note that $B_t^M=\{t\}$, and $B_t^L=V$, so that

$$\beta_M(B_t^M)=\alpha(B_t^M)=0,\qquad \alpha(B_t^L)=\alpha(V),\qquad \beta_L(B_t^L)\ge 0.$$

This allows to use the following telescoping sum, to which we can apply (4.42):

$$\begin{aligned}\alpha(V)&\le\inf_{t\in\widetilde{V}}\sum_{i=L}^{M-1}\alpha(B_t^i)+\beta_i(B_t^i)-\alpha(B_t^{i+1})-\beta_{i+1}(B_t^{i+1})\\&\le 36\inf_{t\in\widetilde{V}}\Psi_t^L+\tfrac12 6^{-L+2}\\&\le 36\inf_{t\in\widetilde{V}}\Psi_t^L+\tfrac12\alpha(V),\end{aligned}$$

so that

$$\alpha(V)\ \le\ 72\inf_{t\in\widetilde{V}}\Psi_t^L. \tag{4.43}$$

Now we define a family of comparison processes. Let $\{\{\xi_B\}_{B\in\mathcal{B}_i}\}_{i=L+1}^M$ be a collection of i.i.d. standard Gaussian variables, and for $t\in\widetilde{V}$ set $\xi_t^i=\xi_{B_t^i}$. Define, for $t\in\widetilde{V}$ and $L\le k\le M-1$,

$$Y_t^k=\sum_{i=k+1}^M 6^{-i}\,\xi_t^i. \tag{4.44}$$

For $s,t \in \widetilde{V}$ with $d(s,t) > 0$ we have that $6^{-i_o-1} < d(s,t) < 6^{-i_o}$ for some i_o, which puts s and t in the same element of $\mathcal{B}_i$ for all $i \le \max(L, i_o - 2)$ so that

$$\begin{aligned} E(Y_t^L - T_s^L)^2 &= E\Big(\sum_{i=\max(L+1,i_o-1)}^{M-1} 6^{-i}(\xi_t^i - \xi_s^i)\Big)^2 \\ &\le 2\sum_{i=i_o-1}^{\infty} 6^{-2i} \\ &\le \frac{2\cdot 6^4}{35} E(X_t - X_s)^2. \end{aligned} \tag{4.45}$$

Set $\mathcal{C}_k = \{B \cap \widetilde{V} : B \in \mathcal{B}_k\}$. Then, from Lemma 4.10 and the definition of Ψ_t^k, we have that there is a universal constant K such that for $k = M-1$ and for each $B \in \mathcal{C}_k$

$$E \sup_{t\in B} Y_t^k \ \ge\ K \inf_{t\in B} \Psi_t^k. \tag{4.46}$$

If we could show that (4.46) was also true for $k = L$, we would be done, for then by (4.43), (4.45), Theorem 2.8 (the weak version of the Sudakov-Fernique inequality) and the fact that $\widetilde{V} \subseteq V$ we would have that

$$\alpha(V) \ \le\ K E\|Y^L\|_{\widetilde{V}} \ \le\ K' E\|X\|_{\widetilde{V}} \ \le\ E\|X\|_V,$$

as required.

Fortunately, we can get precisely what we need by backward induction. Thus, assume that (4.46) is true for some $L < k \le M-1$. We shall show that it is also true for $k-1$. This is all we need.

Thus, with k fixed, choose a $B \in \mathcal{C}_{k-1}$. Set $\mathcal{C}_k^B = \{C \in \mathcal{C}_k : C \subseteq B\}$, and let Ω_C denote the event that $Y_C > Y_{C'}$ for all $C' \neq C$ in $\mathcal{C}_k^B$. Because of the finiteness of the parameter spaces, there is a well defined random variable $\tau : \Omega \to \mathcal{C}_k^B$ defined by the relationship

$$Y_{\tau(\omega)}^k(\omega) \ =\ \sup_{t\in C} Y_t^k(\omega) \qquad \text{on } \omega \in \Omega_C. \tag{4.47}$$

Now use the independence of the ξ_t^k and (4.46) as follows:

$$\begin{aligned}
E \sup_{t\in B} Y_t^{k-1} &\geq E Y_\tau^{k-1} \\
&= \sum_{C\in C_k^B} E\left\{ I_{\Omega_C} \left(\sup_{t\in C} Y_t^k + 6^{-k}\xi_C \right) \right\} \\
&= \frac{1}{|C_k^B|} \sum_{C\in C_k^B} E \sup_{t\in C} Y_t^k + 6^{-k} E \sup_{C\in C_k^B} \xi_C \\
&\geq \frac{1}{|C_k^B|} \sum_{C\in C_k^B} \inf_{t\in C} \Psi_t^k + K6^{-k} \sqrt{\log |C_k^B|} \\
&\geq K \inf_{t\in B} \Psi_t^{k-1},
\end{aligned}$$

which is precisely what we had to show. ■

REMARK: Because this proof has been so very long and convoluted, you should think carefully about precisely what we have needed to make it work. One of the most important steps has been to move the problem from a general parameter space to a finite one. We have used the finiteness a number of times, but never more crucially than in the very last stage, in the definition of the random variable τ at (4.47). It is this step, more than any other, that once more seems to indicate that majorising measures are closely related to the distributions of the positions in T of suprema. Unfortunately, however, the proof relies on the position of the supremum of the comparison process Y, and not that of X itself, so that we cannot obtain this variable in the final result. (Again, I reiterate that in general we have no assurance that such a random variable need be uniquely defined, but that in the case of continuous X Pollard's (1990) result does give us this.) Thus the construction of majorising measures in the most general situations, remains, for the moment, an elusive goal.

4. Entropy.

We have now completed the main results associated with majorising measures. Since this is, however, a somewhat difficult tool to work with, we now return to the use of a somewhat easier, albeit not quite as efficient, concept.

For $\epsilon > 0$, let $N(\epsilon)$ as usual denote the minimal number of d-balls of radius ϵ needed to cover T. Then, as we noted in the Introduction, $H(\epsilon) = \log N(\epsilon)$ is called the (metric) entropy of T. We refer to any result or condition based on N or H as an entropy result/condition.

The fact that entropy results are generally implied by results involving majorising measures is a consequence of the following important lemma.

4.14 LEMMA. *If $\int_0^\infty (\log N(\epsilon))^{\frac{1}{2}} d\epsilon < \infty$, then there exists a majorising measure m and a universal constant K such that*

$$\sup_{t\in T} \int_0^\eta g\big(m(B(t,\epsilon))\big)\, d\epsilon \;<\; K\Big(\eta|\log\eta| + \int_0^\eta \big(\log N(\epsilon)\big)^{\frac{1}{2}} d\epsilon\Big), \tag{4.48}$$

for all $\eta > 0$.

PROOF: As usual, we can and so shall assume that $\mathrm{diam}(T) = 1$. For $n \geq 0$, let $\{A_{n,1}, \ldots, A_{n,N(2^{-n})}\}$ be a minimal family of d-balls of radius 2^{-n} which cover T. Set

$$B_{n,k} \;=\; A_{n,k} \setminus \bigcup_{j<k} A_{n,j}, \tag{4.49}$$

so that $\mathcal{B}_n = \{B_{n,1}, \ldots, B_{n,N(2^{-n})}\}$ is a partition of T and each B_i is contained in a d-ball of radius 2^{-n}. Define a probability measure m on T by

$$m(A) \;=\; \tfrac{1}{2}\sum_{n=0}^{\infty} 2^{-n}\,\big(N(2^{-n})\big)^{-1} \sum_{B\in\mathcal{B}_n} \chi(A\cap B),$$

where $\chi(A) = 1$ if $A \neq \emptyset$ and $\chi(A) = 0$ otherwise. For every $t \in T$ we have that if $\epsilon \in (2^{-(n+1)}, 2^{-n}]$, then

$$m\big(B(t,\epsilon)\big) \;\geq\; \big(2^{n+1}N(2^{-(n+1)})\big)^{-1}.$$

This implies that, for all $t \in T$ and all $n \geq 0$,

$$\begin{aligned}
\int_0^{2^{-n}} &\Big(\log\big(1/m\big(B(t,\epsilon)\big)\big)\Big)^{\frac{1}{2}} d\epsilon \\
&\leq \sum_{k=n+1}^{\infty} 2^{-k}\Big(\log\big(2^k N(2^{-k})\big)\Big)^{\frac{1}{2}} \\
&\leq \sum_{k=n+1}^{\infty} 2^{-k}\big(k\log 2\big)^{\frac{1}{2}} + 2\int_0^{2^{-n}} \big(\log(N(\epsilon)\big)^{\frac{1}{2}} d\epsilon \\
&\leq (n+2)2^{-n}\sqrt{\log 2} + 2\int_0^{2^{-n}} \big(\log(N(\epsilon)\big)^{\frac{1}{2}} d\epsilon,
\end{aligned}$$

the last line following from a little elementary algebra.

It is now easy to complete the proof, using monotonicity arguments to pass from dyadic η to all η. ∎

An immediate consequence of Theorem 4.1 and Lemma 4.14 is the following:

4.15 COROLLARY. *A sufficient condition for the continuity of a centered Gaussian process on T is that $\int_0^\infty (\log(N(\epsilon))^{\frac{1}{2}} d\epsilon < \infty$. Furthermore, there exists a universal constant K such that*

$$E\|X\| \leq K \int_0^\infty \big(\log N(\epsilon)\big)^{\frac{1}{2}} d\epsilon. \tag{4.50}$$

PROOF: To establish continuity we need only, in the notation of, and by, Theorem 4.5 show that $\gamma_m(\eta) \to 0$ as $\eta \to 0$ for some m. This, however, is immediate from Lemma 4.14 and the finiteness of $\int_0^\infty (\log(N(\epsilon))^{\frac{1}{2}} d\epsilon$.

The inequality (4.50) is an immediate consequence (in the case $D = 1$, which is sufficiently general) of setting $\eta = 1$ in (4.48). ∎

There is also a lower bound to $E\|X\|$ involving entropy conditions, dating back to Sudakov (1971), viz.

$$K \sup_{\epsilon > 0} \epsilon\big(\log N(\epsilon)\big)^{\frac{1}{2}} \leq E\|X\|.$$

Given what we have behind us, this inequality is easy to prove. Details of how to proceed appear in Exercise 4.4.

Since entropy calculations are easier to make than are those based on majorising measures, a natural question to ask is whether or not there is a converse to the above corollary.

When X is *stationary*, the answer is positive, and in 1975 Fernique established the following result for processes on $\Re^d$, $d \geq 1$. We can easily obtain a somewhat more general result from what we have already laboured to prove.

4.16 THEOREM. *Let X be a stationary Gaussian process on a compact group, or a compact subset of an infinite group. Then the following three conditions are equivalent:*

(i) X *is a.s. continuous on* T,

(ii) X *is a.s. bounded on* T,

(iii) $$\int_0^\infty (\log(N(\epsilon)))^{\frac{1}{2}} d\epsilon < \infty.$$

PROOF: That (i) implies (ii) is obvious. That (iii) implies (i) is Corollary 4.15. Thus it suffices to show that (ii) implies (iii), which we shall now do. Note firstly that by Theorem 4.4 we know that

$$\sup_{t\in T} \int_0^\infty g\big(m(B(t,\epsilon))\big) d\epsilon < \infty \tag{4.51}$$

for m either normalised Haar measure on T, in the compact case, or the normalised restriction of infinite Haar measure in the second case. Furthermore, by stationarity, the value of the integral must be independent of t.

For $\epsilon \in (0,1)$ let $M(\epsilon)$ be the maximal number of points $\{t_k\}_{k=1}^{M(\epsilon)}$ in T for which

$$\min_{1 \le j,k \le M(\epsilon)} d(t_j, t_k) > \epsilon.$$

It is easy to check (Exercise 4.3) that

$$N(\epsilon) \le M(\epsilon) \le N(\epsilon/2).$$

Thus, since m has total mass unity, we must have

$$m\big(B(t,\epsilon)\big) \le \big(N(2\epsilon)\big)^{-1}.$$

Consequently, by (4.51) and, in particular, its independence on t

$$\infty > \int_0^\infty g\Big(m\big(B(t,\epsilon)\big)\Big)\, d\epsilon \ge \int_0^\infty \big(\log N(2\epsilon)\big)^{\frac{1}{2}}\, d\epsilon$$
$$= 2\int_0^\infty \big(\log N(\epsilon)\big)^{\frac{1}{2}}\, d\epsilon,$$

which proves the theorem. ∎

That an entropy condition cannot in general provide a necessary and sufficient condition for continuity can be seen by counterexamples. Here is one.

4.17 EXAMPLE. *Let $\{X_n\}_{n\ge 1}$ a sequence of i.i.d. Gaussian variables such that*

$$\sigma_n = \sigma(X_n) = (EX_n^2)^{\frac{1}{2}} = (1+\log n)^{-\frac{1}{2}}. \tag{4.52}$$

Then the sequence $\{X_n\}$ is a.s. bounded but does not satisfy the "finite entropy" condition.

PROOF: Note that for each $n \ge 1$ and λ large enough ($\lambda \ge 2$ will do)

$$P\{X_n \ge \lambda\} \le e^{-\lambda^2/2\sigma_n^2}$$
$$\le e^{-\frac{1}{2}\lambda^2(1+\log n)}$$
$$\le Kn^{-\frac{1}{2}\lambda^2}\, e^{-\frac{1}{2}\lambda^2}.$$

Thus

$$P\{\sup_n |X_n| > \lambda\} = P\{\exists n \ge 1 \colon |X_n| > \lambda\} \tag{4.53}$$
$$\le K e^{-\frac{1}{2}\lambda^2} \sum_{n=1}^{\infty} n^{-\frac{1}{2}\lambda^2}$$
$$\le Ke^{-\frac{1}{2}\lambda^2}.$$

It is now a trivial calculation to check (Borel-Cantelli) that $\sup_n X_n$ is a.s. bounded.

Note that so far we have not had to use the fact that the X_n are independent to prove boundedness of the supremum.

The sequence $\{X_n\}_{n\geq 1}$ does not necessarily give a finite entropy integral however. Now use the fact that the X_n are independent, and take $\epsilon > 0$. Then, for $n < n_\epsilon = \exp(-1 + 1/(2\epsilon^2))$, we have $\sigma(X_n) > \epsilon\sqrt{2}$. Thus,

$$d(n,m) \;>\; 2\epsilon \qquad \text{for all } \; n,m < n_\epsilon .$$

Since this means that n and m cannot belong to the same ϵ ball if $n,m < n_\epsilon$, it follows that $N(\epsilon) \geq n_\epsilon - 1$, so that

$$\inf_{\epsilon > 0} \epsilon (\log N_\epsilon)^{\frac{1}{2}} > 0,$$

and so the metric entropy integral $\int (\log N(\epsilon))^{\frac{1}{2}}\, d\epsilon$ cannot be finite. ∎

Despite the lack of a finite entropy integral, however, there must be an appropriate majorising measure. In this case it is not too hard to find, and an appropriate measure is

$$m(\{n\}) \;=\; \frac{K_\alpha}{n(\log n)^{1+\alpha}} \tag{4.54}$$

for any $\alpha > 0$, where K_α is the appropriate normalising constant. There are many counter-examples of this kind. All are based, as is the above, on being forced to use too many balls to cover parts of the parameter space that are somehow too "thin" for anything important to happen.

One way to get around this, while remaining with metric entropy, is to try to divide the space up in two stages, firstly by looking to see which parts are most likely to be problematic, and only then doing entropy calculations. This is an approach that has turned out to be useful in getting good bounds for the supremum probabilities $P\{\|X\| > \lambda\}$ and we shall employ it to much advantage in the following chapter.

A result based on this approach is the following, which goes part of the way towards closing the gap between the entropy and majorising measure conditions for continuity. It is due to Samorodnitsky (1988). The proof we give is based on a personal communication of Talagrand.

Firstly, note that by Theorem 3.6 it is enough to study a.s. continuity at each point of T individually in order to determine full sample path continuity. Let $t_o \in T$ be fixed, and let $B(t_o, \delta)$ be the d-ball around t_o of radius δ. Furthermore, for each $\eta \in (0, \delta)$, set

$$A(t_o, \delta, \eta) \;=\; \{t \in T\colon \eta < d(t, t_o) \leq \delta\},$$

so that if $\eta < \delta$

$$B(t_o, \delta) \;=\; B(t_o, \eta) \cup A(t_o, \delta, \eta).$$

For a general set $C \subset T$, let $N(C, \epsilon)$ be the minimal number of d-balls of radius ϵ required to cover C. Then the following is true.

4.18 THEOREM. *A sufficient condition for the a.s. continuity of X at the point $t_o \in T$ is the existence of a function $H(s,t)$ satisfying the following two conditions:*

$$\lim_{s \to 0} \int_0^s H^{\frac{1}{2}}(s,t)\, dt \ = \ 0, \tag{4.55}$$

$$\log N\big(B(t_o,\eta,\delta),\epsilon\big) \ \le \ H(\eta,\epsilon). \tag{4.56}$$

for all $\epsilon > 0$ and any $0 < \eta < \delta < \Delta$, Δ some fixed constant.

It is easy to see that Theorem 4.18 gives weaker conditions for continuity than does Corollary 4.15, and there are examples covered by the weaker conditions only. Counterexamples show, however, that even the use of a "two-parameter entropy" cannot give necessary *and* sufficient conditions for continuity.

The proof of this result is a very nice example of how judicious use of Borell's inequality, can easily lead to powerful results.

PROOF OF THEOREM 4.18: Consider the process $Y_t = X_t - X_{t_o}$. Clearly Y induces the same canonical metric on T that X induces, and all entropies remain unchanged.

Let $C_n = \{t \in T\colon 2^{-n} \le d(t,t_o) \le 2^{-n+1}\}$. It clearly suffices to show that for any sequence $\epsilon_n \downarrow 0$ the sequence $P\{\sup_{t\in C_n} Y_t > \epsilon_n\}$ is summable. Note firstly that

$$\int_0^{D(C_n)} \big(\log N(C_n,\epsilon)\big)^{\frac{1}{2}}\, d\epsilon \ \le \ 4\int_0^{2^{-n}} \big(H(2^{-n},\epsilon)\big)^{\frac{1}{2}}\, d\epsilon,$$

by (4.56), so that by Corollary 4.15 and 4.55

$$\lim_{n\to\infty} E \sup_{t\in C_n} Y_t \ = \ 0.$$

Let $\eta_n = E\sup_{t\in C_n} Y_t$. By Borell's inequality

$$P\{\sup_{t\in C_n} Y_t > \epsilon_n\} \ \le \ 2e^{-\frac{1}{2}(\epsilon_n - \eta_n)^2/2^{-n}}.$$

By letting ϵ_n decrease slowly enough we obtain a summable series and so complete the proof. ■

5. Ultrametricity and discrete majorising measures.

If you have ever so much as browsed through the papers of Fernique and Talagrand on Gaussian processes (I would hope that by now you have done more than just browsed) you will have noticed that there is one glaring omission in the current notes. The word "ultrametric", which is so very prevalent there, has not appeared here at all.

A metric space (T, d) is called ultrametric if for all $r, s, t \in T$ we have

$$d(s,t) \leq \max\big(d(s,r),\, d(r,t)\big), \tag{4.57}$$

and a Gaussian process X on a space T is called ultrametric if T, together with the canonical metric generated by X, is ultrametric.

The most important aspect of ultrametricity, from the point of view of Gaussian processes, is that two balls of the same radius are either identical or disjoint. We have already seen two very important examples of ultrametric spaces in this chapter, but before I tell you where they were, let us look at a very general way to construct examples.

Let $\{T_n\}_{n\geq 1}$ be a collection of finite sets, and $\{\pi_m^n : T_n \to T_m\}_{1\leq m\leq n}$ a collection of mappings such that for $m \leq n \leq p$

$$\pi_m^p = \pi_m^n \circ \pi_n^p,$$

and, for $t \in T_n$, $\pi_n^n(t) = t$. If T is the projective limit of the $\{T_n, \pi_m^n\}$, furnished with the natural metric, then T is a compact ultrametric space.

To define a ultrametric Gaussian process on T we argue as follows: Firstly, let π_n denote the projection of T onto T_n. Then to every point $t \in \bigcup_n T_n$ assign a standard Gaussian random variable $\xi(t)$. Take the $\xi(t)$ to be mutually independent. Choose $0 < q < 1$, and for $t \in T$ define

$$X(t) = \sum_{n=1}^{\infty} q^n\, \xi\big(\pi_n(t)\big). \tag{4.58}$$

This process is ultrametric, as you can check for yourself (Exercise 5.1).

Now you should remember where we have used ultrametric spaces in these notes. The proofs of both the upper and lower bounds of Theorem 4.1 relied on comparing a given process to one whose structure was somewhat simpler – simpler enough to permit calculation. In both cases, the processes constructed were ultrametric.

In fact, we could have rewritten our proofs somewhat (thus taking them closer to the style of Talagrand's) by first proving what we needed for ultrametric processes, and afterwards mapping the results across to the general case. (Historically, this is precisely how the decade from the mid 'seventies until the mid 'eighties saw most of the majorising measure results derived.) Whether one works this way, or as we have, is very much a matter of taste. Both routes involve the same amount of work – there are no free lunches here.

To be fair, however, I should point out that there are certain advantages to taking the route of first proving things for ultrametric processes and then generalising. The first is that since the first stage of the proof is easier, it gives you a better chance of checking whether or not a new result may or may not be correct and/or provable. Secondly, it has certain distinct advantages when dealing with non-Gaussian processes.

To see how useful ultrametricity can be in terms of simplifying proofs, you should look at a series of papers by Evans (1988a,b, 1989) that are concerned with the sample path properties and extrema of Gaussian processes indexed by local fields. Although Evans does not use the notion of ultrametricity directly, the structure of his parameter space gives him properties of this kind for free, and so allows for neater proofs and occasionally more powerful results than one can obtain in a general setting. ■

We can now turn to the notion of discrete majorising measures, which, as with ultrametricity, can sometimes save one some work.

If (T, d) is a metric space, then a probability measure m is called a *discrete majorising measure* for T if there exists a countable set $S \subset T$ which supports m and a sequence of mappings $\{\pi_n\}_{n\geq 1}$ from T into S such that

$$d(t, \pi_n(t)) \leq 2^{-n}, \qquad t \in T,\ n \geq 1, \tag{4.59}$$

and

$$\sup_{t\in T} \sum_{n=1}^{\infty} 2^{-n} g(m(\pi_n(t))) < \infty. \tag{4.60}$$

To see an nice example of why it is sometimes easier to use discrete majorising measures than their regular counterparts, you should look at Andersen *et. al.* (1988), which has a detailed and powerful treatment of central limit theorems and laws of the iterated logarithm for empirical processes on very general parameter spaces. At this point I shall content myself by just quoting the main results from there which link the two types of majorising measures. The proof is left as an exercise.

4.19 THEOREM. *If (T, d) has admits a majorising measure μ, then it also admits a discrete majorising measure m. If μ satisfies*

$$\lim_{\delta\to 0} \sup_{t\in T} \int_0^{\delta} g(\mu(B(t,\epsilon)))\, d\epsilon = 0 \tag{4.61}$$

then m can be chosen to satisfy

$$\lim_{k\to\infty} \sup_{t\in T} \sum_{n=k}^{\infty} 2^{-n}\, g(m(\{\pi_n(t)\})) = 0. \tag{4.62}$$

This result, along with what has gone before, is sufficient to establish that essentially all the results that we have stated relating boundedness and continuity of Gaussian processes to majorising measures have a corresponding discrete majorising measure form.

For example, if you look again at the upper bound proof for Theorem 4.1, you will see that we have actually proven there that the comparison process Y of (4.20) is a.s. bounded on T when there exists a discrete majorising measure on T. In fact, we showed that the upper bound in (4.3) can be replaced by

$$E\|X\| \le K \sup_{t\in T} \sum_{n=k}^{\infty} 2^{-n}\, g\big(m(\{\pi_n(t)\})\big). \tag{4.63}$$

You should work through the details yourself in Exercise 5.3. In view of Theorem 4.19, a corresponding lower bound also holds.

Again, for more information on discrete majorising measures see Anderson *et. al.* (1988).

6. Discontinuous Processes.

Up until this point we have concentrated on finding conditions under which a Gaussian process X is continuous, and, once it is known to be continuous, measuring its smoothness in terms of moduli of continuity. It seems only reasonable to make a slight detour to see what happens when X is discontinuous.

Recall first that by the zero-one law of Theorem 3.12, if the probability that X is continuous on T is less than one then this probability must be zero, and so X is actually discontinuous with probability one. It then follows from Theorem 3.6 that, for *each* $t \in T$,

$$P\{\lim_{s\to t} X_s \ne X_t\} = 1.$$

Thus X is discontinuous with probability one at every point in a dense subset of T, and so discontinuous at every point in T.

Like the little girl with the curl in the middle of her forehead, when X is good (continuous) it is very very good, but when it is bad, it is horrid.

My task now, as an adherent of the Gaussian school, is to show you that things aren't quite as bad as they seem, as long as you look at them the right way. The way to do this will be to set $T = [0,1]$ (although any compact $T \subset \Re^k$ would do) and to describe some very elegant results due to Don Geman, written up in an unpublished manuscript, Geman (198?). (If you want details, you will have to write to Geman himself. If enough of you do this, this may convince him to publish the paper, as should have been done long ago!) We start with the concept of Lebesgue density.

Let $A \subseteq [0,1]$ be Lebesgue measurable, and let

$$D_0(A) := \lim_{t \downarrow 0} \frac{\lambda(A \cap (0,t))}{t}$$

be the Lebesgue density of A at zero. (λ is, as usual, Lebesgue measure.) $D_0(A)$ is a measure of how "thick" A is in the vicinity of $t = 0$. The basic idea is going to be that any set that has Lebesgue density zero at 0 is so thin that we are going to be prepared to throw it away when we consider the local behaviour of X near 0.

Thus, if ϕ is a Borel measurable function on $[0,1]$, we define

$$\operatorname{ap}\limsup_{t \downarrow 0} \phi(t) = \inf\{a\colon D_0(\{s\colon \phi(s) > a\}) = 0\},$$

and

$$\operatorname{ap}\liminf_{t \downarrow 0} \phi(t) = \sup\{a\colon D_0(\{s\colon \phi(s) < a\}) = 0\},$$

where the "ap" here is to be read as "approximate". Clearly, if $\operatorname{ap}\liminf_{t\downarrow 0} \phi_t = \operatorname{ap}\limsup_{t\downarrow 0} \phi_t$, we call the resulting number the approximate limit of ϕ at zero, and write it as $\operatorname{ap}\lim_{t\downarrow 0} \phi_t$.

One more definition will suffice for the main result. We shall call a non-negative function $\psi = \psi_\phi$ an *approximate upper function* for ϕ at 0 if

$$\operatorname{ap}\lim_{t \downarrow 0} \frac{|\phi(t) - \phi(0)|}{\psi(t)} \leq 1.$$

Approximate lower functions are similarly defined. Here is Geman's main result.

4.20 THEOREM. *Let X be centered, Gaussian on $[0,1]$, and set*

$$p(u) = (E|X_u - X_0|^2)^{\frac{1}{2}}.$$

If X is stationary, p increasing, and if

$$\lim_{u \downarrow 0} \frac{p(u/|\log u|)}{p(u)} = 1, \tag{4.64}$$

then

$$\operatorname{ap}\limsup_{t \downarrow 0} \frac{|X_t - X_0|}{p(t)\sqrt{\log|\log t|}} \leq 1$$

with probability one. That is, $p(t)\sqrt{\log|\log t|}$ is an a.s. approximate upper function for X at 0.

If X is not stationary, then set

$$\tilde{\rho}(u) = \sup\left\{\frac{E(X_t - X_0)(X_s - X_0)}{p(s)p(t)} : \frac{s}{t} = u, \quad s,t \in (0,1)\right\}, \tag{4.65}$$

and assume that

$$\int_0^1 \frac{ds}{\left(1 - \tilde{\rho}(s)\right)^{\frac{1}{2}}} < \infty. \tag{4.66}$$

Then $p(t)\sqrt{2\log|\log t|}$ is an a.s. approximate upper function for X at 0.

I shall not even attempt to prove this result, other than to note that it is based on the non-probabilistic result that a set A has density 0 at $t = 0$ if, and only if,

$$\int_A \Psi\left(\frac{m(A \cap (0,t))}{t}\right)\frac{dt}{t} < \infty,$$

for a some continuous, strictly increasing Ψ on $[0,1]$ with $\Psi(0) = 0$.

You can find precursors to Theorem 4.20, along with full proofs, in Geman (1977, 1979) and Geman and Zinn (1978), but only Geman (198?) has all the details.

The interest in this theorem is that it covers very many processes that are discontinous in the usual sense, but continuous in the "approximate" sense. In fact, there is a (very reasonable) conjecture that this is true of *all* Gaussian processes on $[0,1]$. Don Geman described it to me as follows: "If, after we have drawn a discontinuous, unbounded, Gaussian sample path on the blackboard, we were to step far enough backwards so that we could no longer see sets of zero Lebesgue density, the sample path would become continuous and bounded".

There are examples in the Exercises.

7. Exercises.

SECTION 4.1:

1.1. Let $m_1, \ldots, m_n$ be a sequence of majorising measures on (T,d), and $\alpha_1, \ldots, \alpha_n$ a sequence of non-negative numbers summing to 1. Show that $\sum_{k=1}^n \alpha_k m_k$ is also a majorising measure. That is, convex combinations of majorising measures yield majorising measures.

SECTION 4.4:

4.1. Let T be an compact metric space. Construct a centered Gaussian process on T whose entropy integral diverges, but whose sample paths are continuous with probability one. (Hint. Use the construction of Theorem 1.6 together with the arguments of §4.4.)

4.2. You are now going to prove Corollary 4.15 directly, following Dudley's (1973) proof.

Let $N(\epsilon)$ be the entropy function for X on T, set $H(\epsilon) = \log N(\epsilon)$ and $f(\eta) = \int_0^\eta (H(\epsilon))^{\frac{1}{2}}\, d\epsilon$. We are going to show that f is a modulus of continuity for X in the canonical metric. This will imply Corollary 4.15.

(i) Define sequences $\delta_n \downarrow 0$ and $\epsilon_n \downarrow 0$ inductively as follows: $\epsilon_1 = 1$. Given $\epsilon_1, \dots, \epsilon_n$, let

$$\begin{aligned} \delta_n &= 2\inf\{\epsilon : H(\epsilon) \le H(\epsilon_n)\}, \\ \epsilon_{n+1} &= \min(\epsilon_n/3, \delta_n). \end{aligned}$$

Show that

$$\tfrac{2}{3}\sum_{m=n}^{\infty} H^{\frac{1}{2}}(\epsilon_m)\epsilon_m \;\le\; f(\epsilon_n) \;\le\; 4\sum_{m=n}^{\infty} H^{\frac{1}{2}}(\epsilon_m)\epsilon_m .$$

(ii) Note that for each $n \ge 1$ there is a set $\mathcal{A}_n \subset T$ such that, for any $t \in T$, $\inf_{s \in \mathcal{A}_n} d(s,t) \le 2\delta_n$. Set $\mathcal{G}_n = \{s - t : s, t \in \mathcal{A}_{n-1} \cup \mathcal{A}_n\}$, and let

$$P_n = P\Big\{ \max_{t \in \mathcal{G}_n} \frac{X(t)}{E[X^2(t)]^{1/2}} \ge 3H^{\frac{1}{2}}(\epsilon_n)\Big\}.$$

By using entropy arguments to bound the number of elements of $\mathcal{G}_n$, along with standard Gaussian inequalities, show that

$$P_n \le 4\exp\{-\tfrac{1}{2}H(\epsilon_n)\}.$$

(iii) For any $t \in T$, set $A_0(t) = t$, and for $n \ge 1$ let $A_n(t) \in \mathcal{A}_n$ be such that $d(t, A_n(t)) \le 2\delta_n$. Write X_t as

$$X_t = \sum_{n=0}^{\infty} \big(X(A_n(t)) - X(A_{n+1}(t))\big),$$

and use this representation plus the results of (i) and (ii) to show that f is a modulus of continuity for X.

4.3. For $\epsilon \in (0,1)$ let $M(\epsilon)$ be the maximal number of points $\{t_k\}_{k=1}^{M(\epsilon)}$ in T for which

$$\min_{1 \le j,k \le M(\epsilon)} d(t_j, t_k) > \epsilon.$$

Show that

$$N(\epsilon) \le M(\epsilon) \le N(\epsilon/2).$$

4.4. To give a direct proof of Sudakov's lower bound

$$K \sup_{\epsilon>0} \epsilon\big(\log N(\epsilon)\big)^{\frac{1}{2}} \le E\|X\|,$$

proceed as follows:

Fix $\epsilon > 0$ and choose a maximal set of points, T_ϵ, say, as in the above exercise. Choose i.i.d. zero mean Gaussian variables $\{Y_n\}_{n=1}^{M(\epsilon)}$ with variance ϵ^2. Apply the Sudakov-Fernique inequality of Theorem 2.8 to compare $E\|X\|_{T_\epsilon}$ to $E\|Y_n\|$. Then bound the latter from below by Lemma 4.10. Now send $\epsilon \to 0$ to get a result involving $E\|X\|_T$. If you did it all properly, you will have proven Sudakov's lower bound.

4.5. Find an example satisfying the entropy condition of Theorem 4.18 but not that of Corollary 4.15.

SECTION 4.5:

5.1. Show that the process defined by (4.58) is in fact ultrametric.

5.2 Prove Theorem 4.19.

(Hint: Use the construction we made in the proof of the upper bounds on $E\|X\|$, defining π_n as there, and setting $\mu_n(t) = m(\pi_n(t))$. This result is due, in this explicit formulation, to Anderson *et. al.* (1988).)

5.3. Prove the upper bound to $E\|X\|$ of (4.63). (Note that since T is finite all integrals involving majorising measures are in fact sums.)

SECTION 4.6:

6.1 Let X_t be stationary on $[0,1]$ and assume $E|X_u - X_0|^2 \sim |\log u|^{-\beta}$, $0 < \beta < \infty$. Show that X is continuous for $\beta > 1$ and discontinuous otherwise. In the former case find the regular modulus of continuity at 0. In both cases, calculate the approximate modulus of continuity.

6.2 Let X_t be an index-β process on $[0,1]$. Show that (4.40) is satisfied for these processes. Calculate regular and approximate upper functions at 0 and compare them.

V. SUPREMA DISTRIBUTIONS

1. Introduction.

We now turn our attention to the distribution of the supremum of a centered, Gaussian X over T, which, as always, we write either as $\|X\|_T$ or simply as $\|X\|$ when there is no danger of confusion. In particular, we shall be interested in the asymptotic behavior of $P\{\|X\| > \lambda\}$ as $\lambda \to \infty$.

As we have already seen, an upper bound to this probability comes from Borell's inequality, which tells us that for all $\lambda > 0$

$$(5.1) \qquad P\{|\,\|X\| - E\|X\|\,| > \lambda\} \;<\; 2e^{\frac{1}{2}\lambda^2/\sigma_T^2},$$

where $\sigma_T^2 = \sup_T EX_t^2$. This implies that for all $\lambda > E\|X\|$

$$(5.2) \qquad P\{\|X\| > \lambda\} \;<\; 2e^{-\frac{1}{2}(\lambda^2 - 2\lambda E\|X\| + \|X\|^2)/\sigma_T^2}.$$

Since Borell's inequality was the key to finding sufficient conditions for Gaussian sample path continuity, that also turned out to be necessary, it seems reasonable to assume that it is close to sharp. Nevertheless, the aim of this chapter is to improve on (5.2) in two directions. Firstly, if we think of the right hand side of (5.2) as $f(\lambda)e^{-\frac{1}{2}\lambda^2/\sigma_T^2}$, then our aim will be to replace this by $\hat{f}(\lambda)e^{-\frac{1}{2}\lambda^2/\sigma_T^2}$, where $\hat{f}(\lambda)$ has a lower order of growth in λ than does the exponentially growing f. In many cases, it is possible to find a polynomial $\hat{f}$.

Having found such $\hat{f}$, we would also like to know if we have found the best possible, and so we shall also be interested in lower bounds for $P\{\|X\| > \lambda\}$. These almost always involve much more work than the upper bounds, so we shall generally suffice with statements without proofs. The situation is highly analagous to the continuity problem: sufficiency was easy, necessity was hard. What is rather interesting in the upper bound proofs, however, is that they proceed via a kind of "leap-froging", in which Borell's inequality is used to improve on itself!

Before we look at the general situation, however, it is worthwhile to look at an optimal situation, in which "almost everything" is known, so as to give us an idea as to what sort of results we can hope for in general.

5.1 THEOREM. *Let X be a centered, stationary Gaussian process on $\Re$, with covariance function R satisfying*

$$(5.3) \qquad R(t) \;=\; 1 \;-\; C|t|^\alpha \;+\; o(|t|^\alpha), \qquad \text{as} \quad t \to 0,$$

where $\alpha \in (0,2]$ and C are positive constants. Then for each fixed $h > 0$ such that $\sup_{\epsilon \le t \le h} R(t) < 1$ for all $\epsilon > 0$,

$$(5.4) \qquad \lim_{\lambda \to \infty} \frac{P\{\|X\|_{[0,h]} > \lambda\}}{\lambda^{2/\alpha}\Psi(\lambda)} \;=\; hC^{1/\alpha}H_\alpha,$$

where H_α is a finite constant depending only on α, and, as always,

$$\Psi(\lambda) = (2\pi)^{-1} \int_\lambda^\infty e^{-\frac{1}{2}x^2}\,dx, \tag{5.5}$$

is the probability that a standard normal variable exceeds λ.

This result dates back to Pickands (1969a,b), and you can find a full and detailed proof in Leadbetter, Lindgren and Rootzén (1983). The fact that the proof is 16 pages long indicates that this is not an easy result, and we should not be too hopeful about obtaining extensions for processes on general parameter spaces or for non-stationary processes. (There is an extension to random fields on $\Re^d$, however. See Adler (1981) and references therein.)

Indeed, even in the case treated in Theorem 5.1, the result is not quite as strong as it at first seems. The problem lies in the fact that except for $\alpha = 1$ and $\alpha = 2$ (see Exercises 1.1 and 1.2) it is not known how to calculate the constant H_α in (5.4). (It *is* known that if Y is the non-stationary process on $\Re$ with mean $-|t|^\alpha$ and covariance function $|s|^\alpha + |t|^\alpha - |t-s|^\alpha$ then

$$H_\alpha = \lim_{T\to\infty} T^{-1} \int_{-\infty}^0 e^{-x} P\Big\{ \sup_{0\le t\le T} Y_t > -x \Big\}\,dx,$$

but this does not seem to very instructive, nor of much help in computing H_α.) Aldous (1989), has some non-rigorous bounds on the values of H_α for general α. Most of these arguments are based on his "Poisson clumping heuristic", but they also use Slepian's inequality to interpolate H_α between its known, exact values. For details, turn to Aldous' monograph, which also makes interesting reading for a number of other reasons.

Perhaps the main message of Theorem 5.1 is that although there is clearly very strong structure to asymptotic extrema distributions, and one should be able to do much better than (5.1) and (5.2), we should not spend too much time worrying about constants, since it is unlikely that we shall ever be able to calculate them.

If we now return to the general situation, there are basically three different cases that require study. The first, and in many ways easiest, is when X is non-homogeneous, and there exists a unique point $t_o \in T$ such that $EX_{t_o}^2 = \sup_T EX_t^2$. It turns out that in remarkably many situations

$$\lim_{\lambda\to\infty} \frac{P\{\|X\| > \lambda\}}{\Psi(\lambda/\sigma_T)} = 1. \tag{5.6}$$

In Section 5.3 we shall give a characterization of these cases, due to Talagrand (1988a).

The second case, which is also not too difficult, is when X is either homogeneous or has constant variance. In the purely homogeneous case, and often when X has constant variance, it is sometimes possible to identify a function $M(\lambda)$, which is closely related to the metric entropy function, and a constant K such that

$$(5.7)\qquad K^{-1}M(\lambda/\sigma)\Psi(\lambda/\sigma) \;\le\; P\{\|X\| > \lambda\} \;\le\; KM(\lambda/\sigma)\Psi(\lambda/\sigma),$$

for all sufficiently large λ, where σ^2 is now the constant variance $\sigma^2 = EX_t^2$. (Weber, (1988), has results in this direction, that are of the precise form (5.7) only when $T = [0,1]$ and the covariance function satisfies some side conditions. Patrik Albin has just informed me that he has some very general results – not yet written up – of this form.)

The last case is probably the most interesting, particularly from the point of view of applications to empirical processes, but also seems to be the hardest. It is certainly the case where results are least tight (i.e. the known upper and lower bounds do not always agree). This is the case in which X does not have a constant variance (and, *a fortiori*, is not stationary) and achieves its maximal variance over a comparatively large set. Typical examples are the pinned Brownian sheet on $[0,1]^2$, which achieves its maximal variance of $1/4$ along the line $st(1-st) = 1/4$, or set indexed versions of this process. We shall treat this case in depth in Section 5.4.

We start, however, with some easy calculations.

2. Some Easy Bounds.

As we have noted at least once already, it is a remarkable and very convenient fact that Borell's inequality, together with a simple application of entropy concepts, can be used to improve on Borell's inequality.

To see this, let X be continuous and let $\epsilon > 0$ be arbitrary. In general, ϵ may depend on λ. For $t \in T$, set

$$(5.8)\qquad \mu(t,\epsilon) \;=\; E \sup_{s\in B(t,\epsilon)} X_s,$$

and

$$(5.9)\qquad \mu(\epsilon) \;=\; \sup_{t\in T} \mu(t,\epsilon).$$

Since $N(\epsilon)$ d-balls of radius ϵ cover T, it is an immediate consequence of Borell's inequality that, for $\lambda > \mu(\epsilon)$,

$$(5.10)\qquad P\{\|X\| > \lambda\} \;\le\; 2N(\epsilon)\, e^{-\frac{1}{2}(\lambda-\mu(\epsilon))^2/\sigma_T^2},$$

where, as usual, $\sigma_T^2 = \sup_T EX_t^2$. The following result shows how, with very little work, (5.10) leads to almost the best bounds for $P\{\|X\| > \lambda\}$.

5.2 THEOREM. *If $N(\epsilon) \le K\epsilon^{-\alpha}$, then for all λ sufficiently large,*

$$P\{\|X\| > \lambda\} \le C_\alpha \lambda^{\alpha+1+\eta} \Psi(\lambda/\sigma_T). \tag{5.11}$$

for every $\eta > 0$, where K and $C_\alpha = C(\alpha, K)$ are finite constants.

REMARK: It is possible to replace the condition that λ be sufficiently large by the seemingly weaker requirement that $\lambda > 0$, simply by choosing, if necessary, a larger C_α. We shall do this is in the future. Nevertheless, you should keep in mind the fact that both Theorem 5.2 and those following are really sharpest for $\lambda \to \infty$. (We shall see, however, that in a least one example infinity is reached very quickly, and asymptotic theory is actually good for very moderate values of λ.)

PROOF: To apply (5.10) we must first compute $\mu(\epsilon, t)$. By Corollary 4.15

$$\begin{aligned} \mu(t,\epsilon) &\le C\int_0^\epsilon \left(\log N(\epsilon)\right)^{\frac{1}{2}} d\epsilon \\ &\le C\int_0^\epsilon \left(\log(1/\epsilon)\right)^{\frac{1}{2}} d\epsilon \\ &\le C\epsilon\sqrt{\log(1/\epsilon)}, \end{aligned} \tag{5.12}$$

where $C = C(\alpha)$ may change from line to line. Set $\epsilon = \epsilon(\lambda) = \lambda^{-1}$, choose λ large enough so that $\lambda > C\lambda^{-1}\sqrt{\log\lambda}$ for the C of (5.12), and substitute into (5.10) to obtain

$$\begin{aligned} P\{\|X\| > \lambda\} &\le C_1 \lambda^\alpha e^{-\frac{1}{2}\left(\lambda - C_2\lambda^{-1}\sqrt{\log\lambda}\right)^2/\sigma_T^2} \\ &\le C_3 \lambda^{\alpha+1} e^{C_4\sqrt{\log\lambda}} \Psi(\lambda/\sigma_T), \end{aligned} \tag{5.13}$$

(Recall – (2.1) – that $\Psi(\lambda) \sim \lambda^{-1}e^{-\frac{1}{2}\lambda^2}$.) Since for $\eta > 0$ and λ large enough $e^{C\sqrt{\log\lambda}} < \lambda^\eta$, this completes the proof. ∎

The bound in (5.11) can be improved slightly, to $C(\lambda\log\lambda)^{\alpha+1}\Psi(\lambda/\sigma_T)$, by choosing $\epsilon = \epsilon(\lambda)$ in the above proof to satisfy $\lambda^{-1} = \epsilon(\log(1/\epsilon))^{1/2}$. The gain, however, is rather small, in view of the fact that the following is true:

5.3 THEOREM. *Under the conditions of Theorem 5.2, there exists a finite, positive C_α such that for all $\lambda > 0$*

$$\begin{aligned} C_\alpha^{-1}\lambda^\alpha\left(\log\lambda\right)^{-\alpha/2}\Psi(\lambda/\sigma_T) &\le P\{\|X\| > \lambda\} \\ &\le C_\alpha \lambda^\alpha \Psi(\lambda/\sigma_T). \end{aligned} \tag{5.14}$$

Theorem (5.3) is a special case of Theorem 5.8 of Section 4 below. What is important to note is that the simple proof we applied to prove Theorem 5.2 does not give the sharper upper bound of Theorem 5.3. While this is not surprising, what *is* surprising, however, is the fact that an almost identical, easy, proof *does* give the sharpest upper bound in the following result:

5.4 THEOREM. *If $N(\epsilon) < a\exp(b\epsilon^{-\alpha})$, $0 < \alpha < 2$, then for all $\lambda > 0$,*

$$P\{\|X\| > \lambda\} \leq C_1 \exp\left(C_2\lambda^{2\alpha/(2+\alpha)}\right)\Psi(\lambda/\sigma_T), \tag{5.15}$$

where a, b, C_1, C_2 are finite, positive constants.

REMARK: Note that you cannot set $\alpha = 0$ in this result to recover either Theorem 5.2 or Theorem 5.3. That the upper bound given here is, under mild side conditions, also a lower bound, is shown in Samorodnitsky (1987b, 1990).

PROOF: The proof is identical to that of the previous theorem, with the exception that (5.12) must be replaced by

$$\begin{aligned}\mu(t,\epsilon) &\leq K\int_0^\epsilon \left(\log\left(a\exp(b\epsilon^{-\alpha})\right)\right)^{\frac{1}{2}} d\epsilon \\ &< K\epsilon^{1-\alpha/2},\end{aligned}$$

and the relation between ϵ and λ given by $\epsilon = \lambda^{-1/(1+\alpha/2)}$. The details of the proof are left to you. ■

3. Processes with a Unique Point of Maximal Variance.

Neither Borell's inequality, nor the improvement made on it in the previous section, made any particular use of the fact that the variance of X_t may vary on T. In this section, we make our first step in this direction, with the following elegant result of Talagrand (1988a).

5.5 THEOREM. *Conditions (5.16) and (a)+(b) below are equivalent.*

$$\lim_{\lambda\to\infty} \frac{P\{\|X\| > \lambda\}}{\Psi(\lambda/\sigma_T)} = 1. \tag{5.16}$$

(a) *There exists a unique $t_o \in T$ such that $EX_{t_o}^2 = \sigma_T^2 = \sup_T EX_t^2$.*
(b) *If, for $h > O$, we define*

$$T_h = \{t \in T\colon E(X_t X_{t_o}) \geq \sigma_T^2 - h^2\}$$

then

$$\lim_{h\to 0} h^{-1} E\|X - X_{t_o}\|_{T_h} = 0. \tag{5.17}$$

In view of the results of Chapter 4, condition (5.17) is quite easy to check. Here is an example, treated initially by Berman (1985).

5.6 COROLLARY. *Let X_t, $t \geq 0$ be centered Gaussian with stationary increments and $X_0 = 0$. Assume*

$$p^2(t) \;=\; E(X_{t+s} - X_s)^2 \;=\; EX_t^2$$

is convex, and that

$$\lim_{t\to 0} \frac{p^2(t)}{t} \;=\; 0. \tag{5.18}$$

Then

$$\lim_{\lambda\to\infty} \frac{P\{\|X\|_{[0,T]} > \lambda\}}{\Psi(\lambda/\sigma_T)} \;=\; 1, \tag{5.19}$$

where $\sigma_T^2 = EX_T^2$.

PROOF: Since X has stationary increments, condition (a) of the theorem is clearly satisfied for each $T > 0$ with $t_o = T$. Consider (b). Since

$$EX_T X_t \;=\; \tfrac{1}{2}\big(p^2(T) + p^2(t) - p^2(T-t)\big),$$

it follows that for $t \in T_h$ we have $p^2(t) \geq p^2(T) - 2h^2 = \sigma_T^2 - 2h^2$.

Since $p^2(t)$ is convex, it has a left derivative at each point, and so (draw a picture) for $t \in T_h$ we have $t \geq T - Kh^2$ for some finite K. Thus, since X has stationary increments, to show that (5.17) holds it suffices to show that

$$\lim_{h\to 0} h^{-1} E \sup_{0\leq t\leq h^2} |X_t| \;=\; 0. \tag{5.20}$$

But this follows immediately from an entropy bound such as Corollary 4.15, along with the condition (5.18). ■

PROOF OF THEOREM 5.5: We shall follow our by now standard practice of only proving the theorem in one direction. In this case, however, we are proving it in the harder, albeit more interesting direction; i.e. that (a)+(b) implies (5.16). For the other half of the proof, see Exercise 3.1.

To save a little on notation, set $\sigma_T = 1$. With t_o the unique point of maximal variance, start by setting

$$a(t) \;=\; E(X_t X_{t_o}), \qquad Y_t = X_t - a(t)X_{t_o}.$$

A simple covariance calculation shows that Y_t is independent of X_{t_o}, and since

$$Y_t - (X_t - X_{t_o}) = \big(1 - a(t)\big)X_{t_o},$$

we have, from the definition of T_h, that

$$E\|Y_t - (X_t - X_{t_o})\|_{T_h} \;\leq\; h^2 E|X_{t_o}|.$$

Thus, if (5.17) holds, we also have

$$\lim_{h\to 0} h^{-1} E\|Y\|_{T_h} \;=\; 0 \tag{5.21}$$

Thus we need only show that, under condition (a), (5.21) implies (5.16).

The idea of the proof is as follows: Since the maximal variance is at t_o, it is most likely that if the process achieves a high supremum it will do so close to t_o. Thus we divide the parameter space up into sets close, closer, and even closer, etc. to t_o, treating each one separately, and noting that in each one there is a maximal variance, strictly less than 1, that, by Borell's inequality, governs the distribution of the supremum. The details are as follows:

Take a $\eta > 0$ with $\eta < 1/16$, and $\alpha \in (0, 1/8)$ such that

$$E\|Y\|_{T_h} \;\leq\; \eta h, \qquad \text{for all } h \leq \alpha. \tag{5.22}$$

For $n \geq 0$, define the following subsets of T:

$$A_n \;=\; T_{2^{-n}\alpha} \;=\; \{t\colon a(t) \geq 1 - 2^{-2n}\alpha^2\}, \qquad B_n \;=\; A_n \setminus A_{n+1}.$$

(Draw a picture for $T = [0,1]$, $t_o = 1$, to see what is happenning.)

Note that the A_n are monotonic non-increasing, the B_n disjoint and

$$1 - 2^{-2n}\alpha^2 \;\leq\; a(t) \;\leq\; 1 - 2^{-2(n+1)}\alpha^2, \qquad \text{for all } t \in B_n. \tag{5.23}$$

Since t_o is the unique point of maximal variance

$$\sup\{EX_t^2\colon\, a(t) \leq 1 - \alpha^2\} \;<\; 1.$$

It follows from Borell's inequality that

$$\lim_{\lambda\to\infty} \frac{P\{\sup\{X_t\colon\, a(t) \leq 1 - \alpha^2\} > \lambda\}}{\Psi(\lambda)} \;=\; 0.$$

Thus to prove the theorem we need only show that

$$\limsup_{\lambda\to\infty} \frac{P\{\|X\|_{A_0} > \lambda\}}{\Psi(\lambda)} \;\leq\; 1 + K\eta^2 \tag{5.24}$$

for some universal constant K. (The corresponding lower bound needed to establish (5.16) is, of course, trivial.) To do this, we use the same type of conditioning argument as in the previous two proofs. Using (5.23) we have that for $x > 0$

$$\begin{aligned} P\{\|X\|_{B_n} > \lambda \mid X_{t_o} = x\} &= P\{\|Y_t + a(t)x\|_{B_n} \geq \lambda\} \\ &\leq P\{\|Y\|_{B_n} > \lambda - x(1 - 2^{-2(n+1)}\alpha^2)\}. \end{aligned} \tag{5.25}$$

To bound this, we need to bound $\sup_{B_n} EY_t^2 \leq \sup_{A_n} EY_t^2$. But

$$\begin{aligned} \sup_{A_n} (EY_t^2)^{\frac{1}{2}} &= \sup_{A_n} \sqrt{\pi/2}\,\{E|Y_t|\} \\ &\leq \sqrt{\pi/2}\,E\{\sup_{A_n} |Y_t|\} \\ &\leq \sqrt{\pi/2}\,\{E|Y_{t_o}| + 2E\|Y\|_{A_n}\} \\ &\leq \sqrt{2\pi}\,\eta\alpha\,2^{-n} \\ &\leq \eta\alpha\,2^{-n+2}, \end{aligned}$$

where the first inequality is a property of normal distributions, the second obvious, the third a consequence of Lemma 3.1, the fourth a consequence of (5.22) and the fact that $Y_{t_o} = 0$, and the final trivial.

Applying this, along with (5.22) and Borell's inequality to (5.25), we find that for $x > 0$

$$\begin{aligned} &P\{\|X\|_{B_n} > \lambda | X_{t_o} = x\} \\ &\leq 2\psi\left(\frac{\lambda - x(1 - 2^{-2(n+1)}\alpha^2) - \eta\alpha 2^{-n}}{\eta\alpha 2^{-n+2}}\right) \\ &= 2\psi\left(\frac{\lambda - x(1 - 2^{-2(n+1)}\alpha^2)}{\eta\alpha 2^{-n+2}} - \frac{1}{4}\right), \end{aligned} \tag{5.26}$$

where $\psi(x) = \sqrt{2\pi}\phi(x)$ and, as usual, $\phi(x) = (2\pi)^{-\frac{1}{2}} e^{-\frac{1}{2}x^2}$. Set

$$\begin{aligned} I_1(\lambda) &= \int_0^\lambda P\{\|X\|_{A_o} > \lambda \mid X_{t_o} = x\}\,\phi(x)\,dx, \\ I_2(\lambda) &= \int_{-\infty}^0 P\{\|X\|_{A_o} > \lambda \mid X_{t_o} = x\}\,\phi(x)\,dx, \end{aligned}$$

so that

$$\begin{aligned} P\{\|X\|_{A_o} > \lambda\} &= P\{X_{t_o} > \lambda\} + P\{\|X\|_{A_o} > \lambda,\, X_{t_o} < \lambda\} \\ &= \Psi(\lambda) + I_1(\lambda) + I_2(\lambda). \end{aligned} \tag{5.27}$$

Treating the easier term first, take $x \le 0$ and note that then

$$P\{\|X\|_{A_o} > \lambda \mid X_{t_o} = x\} = P\{\sup_{A_o}(Y_t + a(t)x) \ge \lambda\} \le P\{\|Y\|_{t_o} \ge \lambda\}.$$

Thus $I_2(\lambda) \le P\{\|Y\|_{A_o} \ge \lambda\}$. Since $\sup_{A_o} EY_t^2 < 1$, Borell's inequality gives us that $\lim_{\lambda\to\infty} I_2(\lambda)/\Psi(\lambda) = 0$. Thus it remains only to treat $I_1(\lambda)$.

Since $A_0 = \cup_{n\ge 0} B_n$, it follows from (5.26) that

$$\begin{aligned} I_1(\lambda) &\le \sum_{n=0}^{\infty} \int_0^{\lambda} P\{\|X\|_{B_n} > \lambda \mid X_{t_o} = x\}\,\phi(x)\,dx \\ &\le \sum_{n=0}^{\infty} \int_0^{\lambda} 2\psi\left(\frac{\lambda - x(1-2^{-2(n+1)}\alpha^2)}{\eta\alpha 2^{-n+2}} - \tfrac{1}{4}\right)\phi(x)\,dx. \end{aligned}$$

Make the change of variable $x = \lambda - z$, set

$$\gamma_n = \frac{2^{-(n+4)}\alpha}{\eta}, \qquad \delta_n = \frac{1-2^{-2(n+1)}\alpha^2}{\eta\alpha 2^{-n+2}},$$

to obtain

$$I_1(\lambda) \le 2\sum_{n=0}^{\infty} \int_0^{\infty} \psi(\gamma_n\lambda + z\delta_n - \tfrac{1}{4})\,\phi(\lambda - z)\,dz. \tag{5.28}$$

Note that for all $\lambda > 1$, and all z there exists a finite K such that

$$\psi(\lambda - z) \le \psi(\lambda)\,e^{\lambda z} \le K\lambda\,\Psi(\lambda)\,e^{\lambda z}.$$

Applying this to (5.28) we find

$$I_1(\lambda) \le 2K\,\Psi(\lambda)\sum_{n=0}^{\infty} \int_{n=0}^{\infty} \lambda\psi(\gamma_n\lambda + z\delta_n - \tfrac{1}{4})\,e^{\lambda z}\,dz. \tag{5.29}$$

But, since $\alpha^2 < 1/8$, $\eta < 1/16$, we have

$$\gamma_n\delta_n = 2^{-6}\left(\frac{1-2^{-2(n+1)}\alpha^2}{\eta^2}\right) \ge \frac{2^{-7}}{\eta^2} \ge 2, \tag{5.30}$$

so that returning to (5.29) we find that, for appropriate constants K,

$$\begin{aligned} I_1(\lambda) &\le K\,\Psi(\lambda)\sum_{n=0}^{\infty} \int_0^{\infty} \lambda \exp\left(-\tfrac{1}{2}(\gamma_n\lambda - \tfrac{1}{4})^2 - \tfrac{1}{2}(\delta_n z - \tfrac{1}{4})^2\right) dz \\ &\le K\,\Psi(\lambda)\sum_{n=0}^{\infty} \lambda\,\delta_n^{-1} \exp\left(-\tfrac{1}{2}(\gamma_n\lambda - \tfrac{1}{4})^2\right). \end{aligned}$$

But, applying (5.30) again yields

$$(5.31) \qquad I_1(\lambda) \le \eta^2 K \Psi(\lambda) \sum_{n=0}^{\infty} \lambda\gamma_n \exp\left(-\tfrac{1}{2}(\gamma_n\lambda - \tfrac{1}{4})^2\right).$$

Since $\gamma_{n+1} = \frac{1}{2}\gamma_n$, it is immediate that for large enough λ the sum is bounded by a universal constant.

Thus, $\lim_{\lambda\to\infty} I_1(\lambda)/\Psi(\lambda) \le \eta^2 K$, as required, and since η was arbitrary, the proof is complete. ∎

If we are prepared to assume a little more, it is possible to do better than Theorem 5.5. In the notation of Theorem 5.5, set

$$L(h) = E \sup_{t\in T_h} \left(X_t - a(t)X_{t_o}\right).$$

Recall that condition (5.17) of the Theorem is equivalent (c.f. (5.21)) to

$$\lim_{h\to 0} h^{-1}L(h) = 0,$$

i.e. $L(h) = o(h)$. Dobric, Marcus and Weber (1989) have shown that if we assume a little more about $L(h)$ we can improve Theorem 5.5 as follows.

5.7 THEOREM. *Suppose there exists a unique $t_o \in T$ such that $EX_{t_o}^2 = 1 = \sup_T EX_t^2$, and two functions ω_1, ω_2, concave for $h \in [0, \bar{h}]$ for some $\bar{h} > 0$ with $\omega_i(0) = 0$. Define*

$$h_i(\lambda) = \sup\left\{h : \frac{\omega_i(h)}{h^2} = \lambda\right\}, \qquad i = 1, 2.$$

Then if

$$\omega_1(h) \le L(h) \le \omega_2(h)$$

and $\sup_{t\in T_h} E[X_t - X_{t_o}E(X_tX_{t_o})]^2 < (2-\epsilon)h^2$ for $h \in [0,\bar{h}]$ for some $\epsilon > 0$, there exist constants C_1 and C_2 such that for all λ large enough

$$e^{C_1\lambda\omega_1(h_1(\lambda))} \le \frac{P\{\|X\| > \lambda\}}{\Psi(\lambda)} \le e^{C_2\lambda\omega_2(h_2(\lambda))}.$$

If, for example,

$$\limsup_{h\to 0} \frac{L(h)}{\omega_1(h)} \ge 1,$$

then, for all $\epsilon > 0$,

$$\limsup_{\lambda\to\infty} \frac{P\{\|X\| > \lambda\}}{\Psi(\lambda)\exp(k_1(1-\epsilon)\lambda\omega_1(h_1(\lambda)))} \ge 1.$$

You can find a proof and applications of this result in Dobric, Marcus and Weber (1988). I brought it here to show you two things: Firstly, one *can* do better than Theorem 5.5. Secondly, in terms of the simplicity of the result, one pays a lot the improvement.

4. General Bounds.

At the time of writing, this is the only remaining area in the general theory of Gaussian extremes (over bounded T) that is not yet fully "tidied up" under the influence of Borell's inequality. Thus, we shall suffice with a statement of the main result, without proof.

Nevertheless, the result and proof of the previous section give us a strong indication of how the proof must work. There, we treated processes with a unique maximal variance, and the proof relied on studying just how fast the variance decayed as one moved away from that point. In the general case, there will be a subset $T_{\max} \subset T$ where the maximal variance is achieved, and one has to study two things: the "size" of $T_{\max}$ (e.g. as measured in terms of metric entropy) and how rapidly EX_t^2 decays as we move out of $T_{\max}$.

To quantify this, define the following, for $\delta, \delta_1, \delta_2, \epsilon > 0$:

$$T_\delta^+ = \{t \in T : EX_t^2 > \delta\} \qquad T_\delta^- = \{t \in T : EX_t^2 \leq \delta\}$$

$$N^+(\delta, \epsilon) = N(T_\delta^+, \epsilon), \qquad N^-(\delta, \epsilon) = N(T_\delta^-, \epsilon)$$

$$N(\delta_1, \delta_2, \epsilon) = N(T_{\delta_1}^+ \cap T_{\delta_2}^-, \epsilon) \quad 0 < \delta_1 < \delta_2,$$

where $N(T', \epsilon)$ is the minimum number of d-balls of radius ϵ required to cover $T' \subseteq T$.

It is the function $N(\delta_1, \delta_2, \epsilon)$ that is crucial in measuring the non-stationary of X on T. Here is the main result, due to Samorodnitsky (1987a, 1990) and Adler and Samorodnitsky (1987).

5.8 THEOREM. *Suppose that for finite, positive a, α, β, the following conditions are satisfied for small enough, positive, δ and ϵ:*

$$a^{-1}\epsilon^{-\alpha}\delta^{\beta} \leq N^+(\sigma_T - \delta, \epsilon) \leq a\epsilon^{-\alpha}\delta^{\beta}. \tag{5.32}$$

Then, if $\alpha \geq 2\beta$, there exists a finite, positive K such that for all $\lambda > 0$

$$\begin{aligned} K^{-1}\lambda^{\alpha-2\beta}(\log\lambda)^{-\alpha/2}\,\Psi(\lambda/\sigma_T) &\\ \leq P\{\|X\| > \lambda\} &\\ \leq K\lambda^{\alpha-2\beta}(\log\lambda)^{\beta}\,\Psi(\lambda/\sigma_T). & \end{aligned} \tag{5.33}$$

If $\alpha < 2\beta$, then

$$K^{-1}\Psi(\lambda/\sigma_T) \leq P\{\|X\| > \lambda\} \leq K\Psi(\lambda/\sigma_T). \tag{5.34}$$

Under the additional assumption

$$N(\delta_1,\delta_2,\epsilon) \le C\epsilon^{-\alpha}(\delta_2-\delta_1)^{\beta}, \tag{5.35}$$

the upper bound in (5.33) holds without the logarithmic term.

This result, of course, treats only polynomial entropy functions of the type appearing in Theorems 5.2 and 5.3. (Note, in fact, that setting $\beta = 0$ above in fact gives Theorem 5.3.) The reason that we have not added the exponential entropy case to theorem 5.8 is due to the fact, already noted, that the easy bound of Theorem 5.4 is actually the best possible.

We shall conclude this section with two applications of Theorem 5.8. The first, which treats a non-stationary process on $[0,1]$, is designed to show how sample roughness and non-stationarity interact to determine the distribution of $\|X\|$. (You can get a less interesting result for stationary processes here by setting the parameter β to zero.) The second shows how to use the general theory to do an empirical process problem.

5.9 EXAMPLE. *Let X be stationary, centered on $[0,1]$ with covariance function R_X satisfying*

$$a_0 t^{\alpha} \le 1 - R_X(t) \le a_1 t^{\alpha}, \tag{5.36}$$

for all $t \in [0,\gamma]$ and a_0, a_1, α and γ positive. Let $\sigma(t)$ be positive, continuous, and monotonic increasing on $[0,1]$ such that

$$b_0|t-s|^{\beta} \le |\sigma(t)-\sigma(s)| \le b_1|t-s|^{\beta} \tag{5.37}$$

for all $s,t \in [0,1]$ and b_0, b_1 and β positive, and define a new, nonstationary, process Y by

$$Y(t) = \sigma(t)X(t), \qquad t \in [0,1]. \tag{5.38}$$

Finally, let σ, without a parameter, denote $\sigma(1)$. Then, if $0 < \beta \le \alpha$, we have that for sufficiently large λ there exists a finite $C > 0$ such that

$$C^{-1}\lambda^{-1}e^{-\frac{1}{2}\lambda^2/\sigma^2} \le P\{\sup_{t\in[0,1]} Y(t) > \lambda\} \le C\lambda^{-1}e^{-\frac{1}{2}\lambda^2/\sigma^2}. \tag{5.39}$$

If, however, $0 < \alpha < \beta$, then

$$\begin{aligned} C^{-1}\lambda^{-1-\frac{2}{\beta}+\frac{2}{\alpha}}(\log\lambda)^{-\frac{1}{\alpha}}e^{-\frac{1}{2}\lambda^2/\sigma^2} &\le P\{\sup_{t\in[0,1]} Y(t) > \lambda\} \\ &\le C\lambda^{-1-\frac{2}{\beta}+\frac{2}{\alpha}}e^{-\frac{1}{2}\lambda^2/\sigma^2}. \end{aligned} \tag{5.40}$$

PROOF: Clearly everything hinges on finding good upper and lower bounds for $N(\delta_1, \delta_2, \epsilon) = N(T^+_{\delta_1} \cap T^-_{\delta_2}, \epsilon)$, $0 < \delta_1 < \delta_2$, so that we can establish (5.32) and (5.35). We shall show how to get an upper bound. That the same expression also serves as a lower bound is left as an exercise.

Note first that for $\sigma(0) \le \delta_1 \le \delta_2$,

$$\begin{aligned} T^+_{\delta_1} \cap T^-_{\delta_2} &= \{t \in [0,1]: \delta_1 \le \sigma_t^2 < \delta_2\} \\ &= I(\delta_1, \delta_2) \end{aligned}$$

Note that, in view of (5.37), the length of the interval $I(\delta_1, \delta_2)$ lies between $((\delta_2 - \delta_1)/b_2)^{1/2}$ and $((\delta_2 - \delta_1)/b_1)^{1/2}$. Furthermore, in view of the definition (5.38),

$$\begin{aligned} E(Y_t - Y_s)^2 &= (\sigma_t - \sigma_s)^2 + 2\sigma_t\sigma_s(1 - R_X(t-s)) \\ &\le b_1^2|t-s|^{2\beta} + 2\sigma^2 a_1|t-s|^\alpha, \end{aligned} \tag{5.41}$$

the inequality holding for all $|t-s| < \gamma$, by (5.38).

Moving now to the canonical metric on $[0,1]$ in order to calculate entropies, it follows that, if $0 < \alpha < 2\beta$, then (5.41) implies

$$d(s,t) \le C|t-s|^{\alpha/2}.$$

Combining this with the comments above on the length of $I(\delta_1, \delta_2)$, we have that

$$N(\delta_1, \delta_2, \epsilon) \le C\epsilon^{-2/\alpha}(\delta_2 - \delta_1)^{1/2}. \tag{5.42}$$

This is enough to establish the upper bounds in both (5.32) and (5.35) and so yield, by the Theorem, and Exercise 4.1, (5.39) and (5.40) for the case $\alpha < 2\beta$.

The case $\alpha \ge 2\beta$ is similarly treated, and leads to (5.39). This completes the proof. ∎

Our second example deals with the set indexed, pinned, Brownian sheet $\mathring{W}$ on the collection

$$\mathcal{A}_k = \{[s,t]: s,t \in [0,1]^k\}$$

of intervals in the k-dimensional unit square. Recall that if we take a general reference (probability) measure ν on $[0,1]^k$ then $\mathring{W}$ is the centered Gaussian process with covariance function

$$E\mathring{W}(A)\mathring{W}(B) = \nu(A \cap B) - \nu(A)\nu(B).$$

The result is

5.10 EXAMPLE. *Assume that the measure ν has a bounded density f that is everywhere positive. Then, with the above notation, there exists a finite $C > 0$ such that for large enough λ*

$$C^{-1}\lambda^{2(2k-1)}(\log\lambda)^{-2k}e^{-2\lambda^2} \le P\{\sup_{A\in\mathcal{A}_k} \mathring{W}(A) > \lambda\}$$

$$\le C\lambda^{2(2k-1)}e^{-2\lambda^2}. \tag{5.43}$$

REMARK: The condition on ν can be relaxed to demanding that ν has a strictly positive density on some interval $[s-\epsilon, t+\epsilon]$ $(s,t,\epsilon \in [0,1]^k)$ for which $\nu([s,t]) \ge \frac{1}{2}$, or, indeed, even further to those used by Adler and Brown (1986) in obtaining corresponding bounds (but without the logarithmic term) for $\mathring{W}$ over the index set $\{[0,t]: t \in [0,1]^k\}$. See Samorodnitsky (1987a, 1990) for details.

PROOF: A detailed proof involves an unentertaining amount of algebra, so we shall suffice with only an outline.

For $0 \le \delta_1 < \delta_2 \le \frac{1}{2}$,

$$T^+_{\delta_1} \cap T^-_{\delta_2} = \{A \in \mathcal{A}_k : \delta_1 \le \nu(A) - \nu^2(A) \le \delta_2\}$$
$$= I^-_{\delta_1,\delta_2} \cup I^+_{\delta_1,\delta_2},$$

where

$$I^-_{\delta_1,\delta_2} := \{A \in \mathcal{A}_k : \tfrac{1}{2} - (\tfrac{1}{4} - \delta_1)^{\frac{1}{2}} \le \nu(A) \le \tfrac{1}{2} - (\tfrac{1}{4} - \delta_2)^{\frac{1}{2}}\}$$
$$I^+_{\delta_1,\delta_2} := \{A \in \mathcal{A}_k : \tfrac{1}{2} + (\tfrac{1}{4} - \delta_1)^{\frac{1}{2}} \le \nu(A) \le \tfrac{1}{2} + (\tfrac{1}{4} - \delta_2)^{\frac{1}{2}}\}$$

Consider approximating the sets in I^- in terms of the canonical distance, in order to find an upper bound for the entropy function. (I^+ is similar. The lower bound is left to you. It is a little harder.) Choose as approximating sets k-dimensional intervals whose endpoints sit on the points of the lattice

$$L^k_\epsilon = \{t \in [0,1]^k : t_i = n_i\epsilon^2,\ n_i \in (0,1,\ldots,[\epsilon^{-2}]),\ i = 1,\ldots,k\}.$$

What now has to be shown is that one needs $O(\epsilon^{-4k}(\delta_2-\delta_1)^{\frac{1}{2}})$ such intervals to approximate every set in $I^-_{\delta_1,\delta_2}$ in the canonical metric. This, along with a similar bound for $I^-_{\delta_1,\delta_2}$ yields that

$$N(\delta_1,\delta_2,\epsilon) \le C\epsilon^{-4k}(\delta_2-\delta_1)^{\frac{1}{2}},$$

which, in view of Theorem 5.9, is all we need in order to establish (5.43). ∎

Once one has worked one of these examples in detail, it is relatively easy to work others. For example, for the Brownian sheet (based on Lebesgue measure) on $\{\{\text{half planes}\} \cap [0,1]^2\}$ a result similar to (5.43) holds, but with the power of λ reduced to 2 and the power of the logarithm equal to -2. (Samorodnitsky (1987a, 1990).) As I noted way back in the Preface, the monograph by Piterbarg (1988), which I received only recently, has a wealth of information related to the results of this subsection. His results also cover the more classical "extremal theory" of Gaussian processes, that arises when T is replaced by a sequence $\{T_n\}_{n\geq 1}$ of subsets of Euclidean space that grow towards all of $\Re^d$ as $n \to \infty$.

5. The Brownian Sheet on the Unit Square.

After all the general theory of this chapter it is rather natural to ask whether one could not do better, given a particular process, calculating $P\{\|X\| > \lambda\}$ for this process alone. Perhaps, then, it would be possible to identify the elusive constants in the general bounds, or, better still, calculate the precise distribution of the supremum. Sometimes, but rarely, this is possible. For example, there are only six different stationary Gaussian processes – i.e. six different covariance functions – on $\Re$ for which the precise distribution of the supremum over finite intervals is known.

One is the simple cosine process of Exercise 1.1, and another the process with triangular covariance function (Exercise 1.2). The most important is the Ornstein-Uhlenbeck process with covariance $\exp(|t|)$, and the other three are somehow related to either this or the triangular covariance process. In all of these six cases, however, the derivation of the distribution of the maximum is based not on general Gaussian techniques but either on the simplistic nature of the process (as in the cosine process) or on Markov methods. (The Ornstein-Uhlenbeck process is, of course, Markovian. The triangular covariance process is pseudo-Markovian. See Adler (1981) for definitions and references.) Thus these six "success stories" are not really part of the Gaussian framework.

Within the Gaussian framework, we noted already at the beginning of this chapter that pre-entropy results identify the constants appearing in the asymptotic bounds when $T = [0,1]$ and $R(s-t) \sim |t-s|^\alpha$ as $|t-s| \to 0$. However, even there it is not known how to obtain their precise *numerical* values. In general, even establishing the fact that such constants exist is an impossible task.

Nevertheless, in this section I want to show you how to calculate the constants for one very special process, the pinned Brownian sheet on $[0,1]^2$. There are two reasons for this. Firstly, the result has importance in bivariate Kolmogorov-Smirnov testing. Secondly, it is my favourite result in the entire theory of Gaussian random fields (i.e. processes defined on $\Re^k$, $k > 1$). The result and proof is due to Goodman (1975). A related upper bound is due to Cabaña and Wschebor (1981). It is harder to obtain and not as neat. A higher dimensional version, and a linkage of the result to Kolmogorov-Smirnov tests for empirical processes is given in Adler and Brown (1986). The only other example that I know of for which a similar argument works is in Adler (1984), where the random field on $\Re^2$ with covariance function

$$R(s,t) = \prod_{i=1}^{2} \max\left(0, 1 - |t_i - s_i|\right)$$

is treated. (It is probably worthwhile reiterating at this point that one of the reasons that special results of this type are of interest is that information on a specific process, along with Slepian's inequality, yields information on many others. See, for example, applications of this idea in Orsingher (1987) and Orsingher and Bassan (1988).)

Finally, it is important to note that the reasons that we can get results for the Brownian sheet are more related to its Markovian nature than its Gaussian properties. Nevertheless, it is a lovely calculation, particularly when one recalls its history, and the fact that it grew out of the theory of Banach space valued processes. It is rare indeed that such abstract theory can be used to carry out a calculation that provides useful numbers.

Once again, recall that the Brownian sheet on $[0,1]^2$ is the centered Gaussian process with covariance

$$EW_{st}W_{s't'} = (s \wedge s') \cdot (t \wedge t') \tag{5.44}$$

and the pinned sheet, a version of which is given by

$$\mathring{W}_{st} = W_{st} - stW_{11}. \tag{5.45}$$

has covariance function

$$E\{\mathring{W}_{st}\mathring{W}_{uv}\} = su(1 - tv), \qquad (s,t),(u,v) \in [0,1]^2. \tag{5.46}$$

5.11 THEOREM. *If $\mathring{W}$ is the pinned Brownian sheet on $[0,1]^2$, then for all $\lambda > 0$*

$$P\{\sup_{[0,1]^2} \mathring{W}_{st} > \lambda\} > (1 + 2\lambda^2)e^{-2\lambda^2}. \tag{5.47}$$

The lower bound is, from the point of view of applications, the important one. For $\lambda > 1.5$, it can be shown that it underestimates the true probability by no more than .01, and that it actually gets tighter as $\lambda \to \infty$. (See Theorem 5.12 and the table following it.)

PROOF: The trick is to turn W_{st} from a real-valued, two-parameter process, to a Banach space valued, single parameter, Markov process, by, for each $t \in [0,1]$ letting W_t be the element of $C[0,1]$ defined pointwise by

$$W_t(s) = W_{st}, \qquad 0 \le s \le 1. \tag{5.48}$$

The proof will then rely on the fact that

$$\sup_{st} W_{st} = \sup_t(\sup_s W_t(s)). \tag{5.49}$$

We need to collect a few facts, however, before we can exploit this equivalence fully.

The first is that W_t is, in fact, a $C[0,1]$ valued Markov process. Since this statement is beyond the general interest of these notes, we shall assume it is correct. (If you are uncomfortable with such processes, you can take $0 \le s_1 \le ... \le s_k \le 1$, and work with the $\Re^k$-valued process $W_t^{(k)} = (W(s_1,t)...,W(s_k,t))$ in all that follows and, ultimately, send $k \to \infty$. There is no difficulty checking the Markovian nature of $W^{(k)}$).

The second fact is that one way to obtain $\mathring{W}$ from W is simply to condition W on taking the value zero at the point (1,1). This follows by comparing covariance functions.

Thirdly, since W_t is a well defined Markov process, it makes sense to condition W_t, $0 \le t < 1$ on W_1. A surprising fact, which is what makes the whole proof work, is that $W_t(s)$, conditioned on $W_1(s)$, is independent of $W_1(u)$ for all $u \ne s$. To see this, note that for any $0 \le s,u \le 1$, $0 \le t \le 1$

$$\begin{aligned} E\{(W_t(s) - tW_1(s))W_1(u)\} &= E\{(W_{st} - tW_{s1})W_{u1}\} \\ &= (s \wedge u)(t \wedge 1) - t(s \wedge u)(1 \wedge 1) \\ &= 0. \end{aligned} \tag{5.50}$$

Thus $E\{W_t(s) \mid W_1\} = tW_1(s)$, and our claim is established.

(It is worthwhile reiterating that (5.50) is a *very* special relationship, that will not work, for example, if the Gaussian white noise that W is based on is defined relative to any measure other than Lebesgue.)

The fourth, and final, fact, is that the process W_1, conditional on $W_1(1) = 0$, is a standard Brownian bridge (i.e. the one-dimensional Brownian sheet).

We can now start the calculation:

$$\begin{aligned} P\{\sup_{st} \mathring{W}_{st} < \lambda\} &= P\{\sup_{st} W(s,t) < \lambda \mid W(1,1) = 0\} \\ &\le \inf_s P\{\sup_t W_t(s)) < \lambda \mid W(1,1) = 0\} \\ &= \int_{\Omega^*} \inf_s P\{\sup_t W_t(s) < \lambda \mid W_1 = w\}\, dP^*(\omega), \end{aligned}$$

where P^* is the probability measure on $\Omega^* = \{f \in C[0,1]: f(0) = f(1) = 0\}$ induced by the standard Brownian bridge. (If the last equality bothers you – and it should – see Goodman's original proof for technical elucidation.)

By the third fact established above, we now have that

$$P\{\sup_{st} \mathring{W}_{st} < \lambda\}$$
$$\leq \int_{\Omega^*} \inf_s P\{\sup_t W_t(s) < \lambda \mid W_1(s) = \omega(s)\} \, dP^*(\omega). \tag{5.51}$$

But for each s, the real valued process $\{W_t(s)\}_{t \geq 0}$ is just a scaled Brownian motion, and the conditional probability that $W_t(s) < \lambda$ for all $0 \leq t \leq 1$, given the value of $W_1(s)$, is well known, as a property of Brownian motion. In fact (e.g. Feller (1971), page 341) we have

$$P\{\sup_t W_t(s) \;<\; \lambda \mid W_1(s) = w(s)\} \;=\; 1 - \exp\left(-2\lambda \frac{(\lambda - w(s))}{s}\right).$$

Substituting into (5.51) now gives

$$P\{\sup_{st} \mathring{W}_{st} < \lambda\} \;\leq\; \int_{\Omega^*} \left\{1 - \exp\left(2\lambda \sup_s \frac{(w(s) - \lambda)^-}{s}\right)\right\} dP^*(w). \tag{5.52}$$

However, the distribution of $Z = \sup_s\{(w(s) - \lambda)/s\}$ when ω is a Brownian bridge is also known, and has density

$$p(z) \;=\; \begin{cases} \lambda e^{-2\lambda(z+\lambda)} & z > -\lambda, \\ 0 & z < -\lambda. \end{cases}$$

Substituting into (5.52) gives

$$\begin{aligned} P\{\sup_{st} \mathring{W}_{st} < \lambda\} \;&\leq\; \int_{-\lambda}^{0} (1 - e^{2\lambda z})\, 2\lambda\, e^{-2(z+\lambda)\lambda}\, dz \\ &\leq 1 - (1 + 2\lambda^2)\, e^{-2\lambda^2}. \end{aligned}$$

Consequently,

$$P\{\sup_{st} \mathring{W}_{st} > \lambda\} \;\geq\; (1 + 2\lambda^2)\, e^{-2\lambda^2},$$

as required. ■

While it is true that Theorem 5.11 is one of my favourite results, I should point out that one can actually do considerably better. Furthermore, whereas the proof given above relies on Markov and Banach space techniques, the better result is a truly Gaussian one. That is, the proof of the result below proceeds by noting that the largest contribution to the probability of the pinned Brownian sheet ever crossing a high level comes from that part of $[0,1]^2$ where the marginal probability of being above the level is highest; i.e. where the variance is highest. This region can then be carefully broken up into smaller pieces, and each one studied independently. This, of course, is precisely the technique that we have employed throughout this chapter. However, when carefully applied to a particular process, it is possible to keep track of all the constants involved, and Hogan and Seigmund (1986) (to whom you should turn for a proof) have thus managed to prove the following:

5.12 THEOREM. *If $\mathring{W}$ is the pinned Brownian sheet on $[0,1]^2$, then as $\lambda \to \infty$*

$$P\{\sup_{[0,1]^2} \mathring{W}_{st} > \lambda\} \sim 4\log 2\, \lambda^2 e^{-2\lambda^2}. \tag{5.53}$$

Note that in one sense, Theorem 5.12 is a little weaker than Theorem 5.11, as the latter gives an inequality for *all* $\lambda > 0$. It is interesting, nevertheless, to compare the two numerically, and so note that "∞" is a lot closer to zero than one might initially expect.

The following table does this, giving $P\{\sup_{[0,1]^2} \mathring{W}_{st} > \lambda\}$ for a range of λ, according to formulae (5.47), (5.53) and an approximation based on 400 simulations of a discrete approximation to a pinned Brownian sheet on a 100×100 grid on $[0,1]^2$ (graciously provided by Ron Pyke).

PINNED BROWNIAN SHEET TAIL PROBABILITIES			
λ	Lower bound	Asymptotic	Simulation
0.25	.9928	.1529	.9883
0.50	.9098	.4204	.9226
0.75	.6899	.5063	.7133
1.00	.4060	.3752	.4191
1.25	.1812	.1903	.1737
1.50	.0611	.0693	.0503
1.75	.0156	.0186	.0116
2.00	.0030	.0037	.0000
2.25	.0004	.0006	.0000

If you have bothered to read these figures then you are enough of a statistician to know that the last digit in the simulation results is totally spurious, and that very little faith should be had in even the third.

A little thought also shows that at the extreme tail the simulation figures should give lower probabilities than the true ones, since an approximation to $\mathring{W}$ on a finite grid will have a stochastically smaller supremum than $\mathring{W}$ itself.

Nevertheless, the overall agreement, at least for $\lambda \geq 1.25$ is surprisingly good, so that one is rather tempted to claim that, with an error of only 1%, $\infty \approx 2.5\sigma$! (Since $\sigma_T^2 = 1/4$.) In fact, experience, based both on numerical approximations and simulation indicates that this approximation is good for a wide variety of extremal results. ∎

There is, of course, one thing wrong with both of the above Theorems: both refer only to the Brownian sheet based on the uniform measure on $[0,1]^2$. With only a little work, and an application of Slepian's inequality, one can do a lot better.

With a slight change of notation, so that the proof below will be easier to read, let F be a continuous distribution function on $\Re^2$. Write $\mathring{W}_F$ to denote the pinned Brownian sheet based on the measure with distribution function F, so that for $s, t \in \Re^2$

$$E\mathring{W}_F(s)\mathring{W}_F(t) = F(s \wedge t) - F(s)F(t), \tag{5.54}$$

where $s \wedge t$ is the coordinatewise minimum $(s_1 \wedge t_1, s_2 \wedge t_2)$.

Furthermore, let G denote the degenerate distribution, uniform on the negatively sloped diagonal $t_1 + t_2 = 1$ in $[0,1]^2$, so that

$$G(t) = (t_1 + t_2 - 1)^+, \qquad t \in [0,1]^2. \tag{5.55}$$

Here is a rather useful and somewhat surprising result:

5.13 THEOREM. *Let F be any continuous distribution function on $\Re^2$. For any $\lambda > 0$*

$$P\{\sup_{t \in \Re^2} \mathring{W}_F(t) > \lambda\} \leq P\{\sup_{t \in [0,1]^2} \mathring{W}_G(t) > \lambda\}. \tag{5.56}$$

Furthermore

$$P\{\sup_{t \in [0,1]^2} \mathring{W}_G(t) > \lambda\} \leq \sum_{n=1}^{\infty} (8n^2\lambda^2 - 2)\, e^{-2n^2\lambda^2}. \tag{5.57}$$

The value in this result lies in its applications to bivariate Kolmogorov-Smirnov tests. You can find more details in Adler and Brown (1986) from where the result comes, as well as in Adler, Brown and Lu (1990).

PROOF: We shall assume, without loss of generality, that F is concentrated on $[0,1]^2$ with uniform marginals. (Why is this true? Also, since this *is* true, why is it not true that we can also assume F to be uniform on $[0,1]^2$? Why *can* we assume this in one dimension?)

To save on space, write I^2 for $[0,1]^2$, and let m be the mapping from I^2 to the upper triangle $\{t \in I^2 : t_1 + t_2 \geq 1\}$, defined by

$$G\big(m(t)\big) = G\big(m_1(t), m_2(t)\big) = F(t), \qquad t \in I^2, \tag{5.58}$$

$$m_2(t) - m_1(t) = t_2 - t_1, \qquad t \in I^2. \tag{5.59}$$

We must check that m is well defined. But this is easy, for by (5.59) we have that we are mapping lines of slope one onto themselves. At the lower left end of each such line $F = G = 0$, and at the upper right end, as each line leaves I^2, the marginal uniformity of F and G ensures that they are again equal. Since both F and G are nondecreasing along such lines, and G is continuous and strictly increasing when it is not zero, it is easy to arrange (5.58) in a unique fashion.

Now consider the processes $\mathring{W}_F$ and $\mathring{W}_G$. We shall compare their suprema via Slepian's inequality, for which we need to compare variances and covariances. Note firstly that, for $t \in I^2$,

$$E\big(\mathring{W}_F(t)\big)^2 = E\big(\mathring{W}_G(m(t)\big)^2, \tag{5.60}$$

a simple consequence of (5.58) and (5.54). We want to show that

$$F(s \wedge t) \geq G\big(m(s) \wedge m(t)\big), \tag{5.61}$$

from which it would follow that

(5.62)
$$\begin{aligned} E\mathring{W}_F(s)\mathring{W}_F(t) &= F(s \wedge t) - F(s)F(t) \\ &\geq G\big(m(s) \wedge m(t)\big) - G\big(m(s)\big)G\big(m(t)\big) \\ &= E\mathring{W}_G\big(m(s)\big)\mathring{W}_G\big(m(t)\big). \end{aligned}$$

This will be enough to prove the first part of the Theorem, viz (5.56), since (5.60) and (5.62) are precisely the ingredients for Slepian's inequality.

Therefore, consider (5.61), and assume that $s < t$ in the sense of coordinatewise ordering. (i.e. $s_1 < t_1$ and $t_1 < t_2$.) Then $F(s \wedge t) = F(s) = G(m(s)) \geq G(m(s) \wedge m(t))$, and so (5.61) holds. The case $t < s$ is clearly identical, so consider now the case $s_1 > t_1$ and $s_2 < t_2$. (The remaining case is handled identically.) Then $s \wedge t = (t_1, s_2)$. Write $w = (m_1(t), m_2(s))$. There are three possible cases to consider: $m(s) \geq w \geq m(t)$, $m(t) \geq w \geq m(s)$, and $w = m(s) \wedge m(t)$. We shall look at the third case only, but you should check that our reasoning is valid for all three. Note (draw a picture!) that

$$\begin{aligned} F(s\wedge t) &= F(t_1,s_2) \\ &\geq \big[F(s)-(s_1-t_1)\big]\vee\big[F(t)-(y_2-s_2)\big] \\ &\qquad\qquad\qquad\text{(by marginal uniformity)} \\ &\geq \tfrac{1}{2}\big\{F(s)+F(t)-\big[(s_1-s_2)-(t_1-t_2)\big]\big\} \\ &= \tfrac{1}{2}\big\{G\big(m(s)\big)+G\big(m(t)\big) \\ &\qquad -\big[\big(m_1(s)-m_2(s)\big)-\big(m_1(t)-m_2(t)\big)\big]\big\} \quad \text{by (5.59), (5.60)} \\ &\geq m_2(s)+m_1(t)-1 \quad \text{by (5.55).} \end{aligned}$$

Hence, if $m_2(s)+m_1(t)-1\geq 0$, then the above yields

$$F(s\wedge t) \;\geq\; G\big(m_1(t),m_2(s)\big). \tag{5.63}$$

On the other hand, if $m_2(s)+m_1(t)-1<0$, then $G(m_1(t),m_2(s))=0$ and so (5.63) is trivially true. Thus, in general,

$$F(s\wedge t)\;\geq\; G\big(m_1(t),m_2(s)\big) \;=\; G(w) \;=\; G\big(m(s)\wedge m(t)\big).$$

From this we immediately obtain (5.61) and so the proof of (5.56).

To complete our proof we must establish (5.57). To this end, we define a two-parameter process $X(t)$ on I^2 by setting

$$X(t_1,t_2) \;=\; \begin{cases} \mathring{W}(t_1)-\mathring{W}(1-t_2) & t_1+t_2-1\geq 0 \\ 0 & t_1+t_2-1\geq 0, \end{cases}$$

where $\mathring{W}$ is the usual single parameter Brownian bridge $\mathring{W}$ on $[0,1]$. Comparison of covariance functions shows that X is a version of $\mathring{W}_G$. Thus

$$\begin{aligned} &P\{\sup_{t\in I^2}\mathring{W}_G(t)>\lambda\} \\ &\quad= P\{\sup\big[\mathring{W}(t_1)-\mathring{W}(t_2): t_1+t_2-1\geq 0\big]>\lambda\} \\ &\quad= P\{\sup\big[\mathring{W}(t)-\mathring{W}(s): 0\leq s<t\leq 1\big]>\lambda\} \\ &\quad\leq P\{\sup\big[\mathring{W}(t)-\mathring{W}(s): s,t\in[0,1]\big]>\lambda\} \\ &\quad= P\Big\{\Big[\sup_{[0,1]}\big(\mathring{W}(t)\big)^+ + \sup_{[0,1]}\big(\mathring{W}(t)\big)^-\Big]>\lambda\Big\}. \end{aligned} \tag{5.64}$$

What we have gained here is that the supremum probability for the two-parameter $\mathring{W}_G$ has been expressed in terms of the single parameter process $\mathring{W}$, and, fortunately for us, Kac, Kiefer and Wolfowitz (1955, equation (4.6)) have shown, via clever use of the reflection principle – which in turn relies on the Markov type properties of the Brownian bridge – that the last probability above is precisely that given by the right hand side of (5.57). This completes our proof. ■

REMARK: For the statistician who is still with us, and who might be interested in good, numerical, bounds, we should note that the one inequality in (5.64) is far from sharp. While it retains the right order of magnitude, a little thought shows that it "costs", roughly, a factor of two. Comparison of the upper bound of Theorem 5.13 and the lower bound of Theorem 5.11 (albeit for a specific process) seems to bear this out, as do simulations in Adler, Brown and Lu (1990), to where, once again, I refer you for details on how to use all of this.

Finally, it is interesting to ask what happens in treating the Brownian sheet in dimensions greater than two. Do similar results hold? Unfortunately, the answer seems to be negative, since there is no distribution in dimensions ≥ 3 which plays the same "extremal" role that G does in two dimensions.

You can look in Adler and Brown (1986) to see what happens in higher dimensions, but all you will find there are special cases of the very general theorems of this chapter. In fact, it was the results obtained there, for particular processes, that got me interested in the general formulation. Since I now seem to have got back to where I started, the cycle is complete, and the time has come to stop writing.

6. Exercises.

SECTION 5.1:

1.1 One of the useful consequences of Theorem 5.1 is that the asymptotic distribution of the maximum of stationary Gaussian processes on $\Re$ depends on the covariance only through the constants C and α in the expansion (5.3). Consequently, if we can find the value of H_α in (5.4) for a particular process X with covariance function satisfying (5.3) for a specific α, we have H_α for all such processes.
(i) Let A be a Rayleigh random variable, with density $f(a) = a\exp(-a^2/2)$, $a \geq 0$, and ϕ uniform on $[0, 2\pi)$, independent of A. Fix $\omega > 0$ and define the "cosine process with angular frequency ω" as

$$X(t) \;=\; A\cos(\omega t \;-\; \phi).$$

Show that X is centered, Gaussian, and stationary. Find its covariance function, and show that it satisfies (5.3) with $\alpha = 2$.
(ii) From elementary geometric considerations, show that for $0 < T < \pi/\omega$ and $\lambda > 0$

$$P\Big\{\sup_{0\leq t\leq T} X(t) > \lambda\Big\} \;=\; \Psi(\lambda) + \frac{\omega T}{2\pi}\exp\big(-\tfrac{1}{2}\lambda^2\big).$$

(iii) Apply (i), (ii), and Theorem 5.1 to show that $H_2 = 1/\sqrt{\pi}$.

1.2 Another specific case in which the distribution of the maximum can be explicitly calculated is the Gaussian process with triangular covariance

$$R(t) = 1 - |t|, \qquad |t| \leq 1.$$

The fact that this process satisfies a pseudo Markov property allowed Slepian (1961) to derive the distribution of the maximum. You might try going over his calculation, and, *en passant*, showing that $H_1 = 1$.

SECTION 5.3:

3.1 We shall now prove the necessity in Theorem 5.5
(i) Show that if (U, V) are centered, jointly Gaussian variables, with $EU^2 = EV^2 = 1$ and $EUV < 1$, then

$$\lim_{\lambda \to \infty} \frac{P\{\max(U,V) \geq \lambda\}}{\Psi(\lambda)} = 2.$$

Hence conclude that if (5.16) holds, then so must condition (a) following it. (That is, if $\|X\|$ behaves like a single Gaussian variable, then X achieves its maximal variance at only one point.)
(ii) Show that for $0 \leq \lambda \leq \nu$,

$$\Psi(\lambda) \geq \Psi(\nu) + \nu(\nu - \lambda)\Psi(\nu).$$

(iii) Assume that (5.16) is in force. By (i), this implies that condition (a) of Theorem 5.5 is also in force. Show now that (5.16) is equivalent to (5.17).
(iv) Complete the proof of necessity by showing that (5.17) implies (5.21). An outline of the main steps is as follows: Choose $\epsilon > 0$ small enough so that

$$P\{\|X\| > \lambda\} \leq (1 + \epsilon^2)\Psi(\lambda),$$

for large enough λ. With $a(t)$ and $Y(t)$ as in the proof of Theorem 5.5, and $T' = \{t \in T\colon a(t) \geq 0\}$, deduce that

$$\int_{-\infty}^{\lambda} \psi(y)\phi(y)\, dy \leq \epsilon^2 \Psi(\lambda),$$

where $\psi(y) = P\{\sup_{t \in T'}(Y(t) + a(t)y) > \lambda\}$, and ϕ, as usual, is a standard Gaussian density.

Now complete the proof. You will need both (ii) and the result of Exercise 1.1 of Chapter 3.

SECTION 5.4:

4.1 Complete the proof of Example 5.8 by showing that the upper bound (5.42) for $N(\delta_1, \delta_2, \epsilon)$ also serves, albeit with a different constant, as a lower bound.

4.2 Fill in all the missing details in the derivation of the upper bound for the entropy in the proof of Example 5.9. If you are feeling truly courageous, try proving that the upper bound also serves as a lower bound, as usual, with a different constant.

VI. AFTERTHOUGHTS

1. Two topics that were left out.

THE $\mathcal{L}^2$ SPACE OF A GAUSSIAN PROCESS: Looking back over what I have written, it seems to me that these notes contain almost everything a beginning researcher needs to know about the mathematical basis of Gaussian processes. Somehow, although my original intention was only to cover problems of continuity and extrema, there is a lot of other material that has found its way into the notes. There is one gaping omission, however, and this relates to the $\mathcal{L}^2$ space associated with a Gaussian process, and its Hilbert space structure.

The reason that this material does not appear here is that it has nothing to do with continuity, boundedness, and suprema distributions, and these have been our main concern. Nevertheless, a word of advice to the teacher or the reader: If, after going through these notes you have some time and energy left, read the first seven chapters of Major's (1981) monograph on multiple Wiener integrals, and follow it up with the elegant and powerful central limit theorem of Dynkin and Mandelbaum (1983). Then you will be ready to start a serious reading of the Gaussian literature in all its breadth and beauty.

ON MARKOV PROCESSES AND THE ISOMORPHISM THEOREM: Markov processes have appeared in these notes only in a very peripheral fashion. This is unfortunate, because it tends to reinforce the "well known fact" that Gaussian and Markovian processes have little to do with one another, and this WKF is taking somewhat of a beating at the moment. To explain why, I want to describe for you a simple version of a theorem due to Dynkin (1983). Then I will explain to you why this theorem is interesting.

Let $\{X_t, t \geq 0\}$ be an $\Re^d$-valued, symmetric, Markov process with stationary transition density $p_t(x,y) = p_t(y,x)$. (We really need certain technical side conditions, but shall leave them out here.) Let ξ be an exponential random variable with mean one, independent of X, which we treat as a death time for X, and let Δ be the "cemetary" state for X so that the "killed version" of X is given by the process

$$\hat{X}_t = \begin{cases} X_t, & t < \xi, \\ \Delta, & t \geq \xi. \end{cases}$$

The killed process $\hat{X}_t$ is still a Markov process, with transition density

$e^{-t}p_t(x,y)$, $x,y \in R^d$, and Green's function $g(x,y)$ defined by

$$g(x,y) = \int_0^\infty e^{-t} p_t(x,y)\, dt. \tag{6.1}$$

We shall assume throughout that $0 < g(x,x) < \infty$ for all x.

For each $x_0 \in R^d$ we define the probability P_{x_0} on the space of paths of X, augmented by Δ, via the finite-dimensional distributions

$$(6.2) \quad P_{x_0}\left\{\hat{X}_{t_1} \in B_1, \ldots, \hat{X}_{t_k} \in B_k\right\}$$
$$= \frac{1}{g(x_0,x_0)} \int_{B_1} \cdots \int_{B_k} e^{-t_k} p_{t_1}(x_0,x_1) p_{t_2-t_1}(x_1,x_2) \ldots p_{t_k-t_{k-1}}(x_{k-1},x_k)$$
$$\times\, g(x_k,x_0)\, dx_1 \ldots dx_k$$

for $t_1, \ldots t_k > 0$ and Borel sets B_i. Properly formulated $P_{x_0}\{\cdot\}$ can be interpreted as $P\{\cdot \mid \hat{X}_0 = x_0, \hat{X}_{\xi-} = x_0\}$, so that P_{x_0} describes a process starting and finishing (after an exponential killing time) at x_0. (Note that at time ξ the path takes the value Δ. Therefore, $P_{x_0}\{\hat{X}_{t_1} \in R^d, \ldots, \hat{X}_{t_k} \in R^d\} = P\{\xi > t_k \mid \hat{X}_0 = x_0, \hat{X}_{\xi-} = x_0\}$.)

We shall be interested in the local time process $\{L_x, x \in R^d\}$ of the killed version $\hat{X}_t$ of X_t, which can be formally expressed as

$$L_x = \int_0^\xi \delta_x(X(t))\, dt, \tag{6.3}$$

where δ_x is the Dirac delta function centered at x. This definition can be made rigorous, as usual, by taking L_x as the Radon-Nikodym derivative of the occupation measure

$$\mu(A) = \int_0^\infty I_A(\hat{X}(t))\, dt,$$

where we follow the convention that for all Borel sets $A \in R^d$, $I_A(\hat{X}_t) = 0$ for $t \geq \xi$. (Another approach is to approximate δ_x by some density function and then pass to the limit.)

To set up Dynkin's theorem, the first thing we have to note is that the

Green's function (6.1) is always positive semi-definite, since

$$\begin{aligned}
&\iint \alpha(x)g(x,y)\alpha(y)\,dxdy \\
&= \int\int\int \alpha(x)e^{-t}p_t(x,y)\alpha(y)\,dxdydt \qquad \text{(by (6.1))} \\
&= \iint\iint \alpha(x)e^{-t/2}p_{t/2}(x,z)\,\alpha(y)e^{-t/2}p_{t/2}(z,y)\,dxdydtdz \\
&\qquad\qquad \text{(by Chapman-Kolmogorov)} \\
&= \iint dzdt\Big\{\int \alpha(x)e^{-t/2}p_{t/2}(z,x)\,dx \int \alpha(y)e^{-t/2}p_{t/2}(z,y)\,dy\Big\} \\
&\qquad\qquad \text{(by symmetry)} \\
&= \iint dzdt\big[E\{\alpha(X_{t/2})|X_0=z\}\big]^2 \\
&\geq 0.
\end{aligned}$$

Thus g can serve as the covariance function of a centered Gaussian process, $\{G(x), x \in R^d\}$, say. Let (Ω, π) denote the probability space of this process. The measure π is, naturally, completely determined by the covariance function $g(x,y)$. Let $L = \{L_x, x \in R^d\}$ denote the local time process defined in (6.3). This process is defined on a probability space (Ω_1, P_{x_0}) where P_{x_0} is given in (6.2). We denote by E_π and $E_{P_{x_0}}$ expectation with respect to the measures π and P_{x_0} respectively. For reasons we shall explain in a moment, the following result is known as "the isomorphism theorem":

6.1 THEOREM (DYNKIN). *Let F be any positive functional on the space of functions from $\Re^d$ to $\Re$ and let $\xi_x = \frac{G^2(x)}{2}$. Then*

$$E_\pi\Big\{F(\xi)\frac{G^2(x_0)}{g(x_0,x_0)}\Big\} = E_\pi E_{P_{x_0}}\{F(\xi+L)\} \tag{6.4}$$

Equivalently, the process $\xi_\cdot(\omega)$ with the measure $\frac{G^2(x_0)}{g(x_0,x_0)}\pi(d\omega)$ and $\xi_\cdot(\omega) + L(\omega_1)$ with the product measure $\pi(d\omega)P_{x_0}(d\omega_1)$ are equal in distribution.

The importance of this Theorem is that it gives us a way of translating probability statements about certain Markov processes (at least those statements that can be phrased in terms of local time) into statements about Gaussian processes, and *vice versa.* Hence the "isomorphism" in the name of the theorem. As an example, note that it is an immediate consequence of the isomorphism theorem that the local time exists as long as $0 < g(x,x) < \infty$, and that it has a continuous version whenever $\{G(x), x \in R^d\}$ has a version with continuous sample paths. This, of course, is a problem that we now know how to solve. Much deeper information about the sample path properties of the local times of Markov processes can also be deduced from known results on Gaussian processes, many of which are extremely difficult to obtain in a purely Markovian framework. For details, see Marcus and Rosen (1990), who are rewriting and extending previous results on Markov local time via Gaussian tools.

In a very clever application of this approach, Raisa Epstein Feldman (1989) has shown how to take all the hard combinatorics out of calculating hitting time probabilities for random walks with exponential holding times by using Gaussian processes. In Feldman (1990) she has shown how to calculate general exit distributions for Lévy processes via Gaussian techniques, and has obtained Wiener-Hopf factorisation results from the same framework.

The relationship between Markov and Gaussian processes goes even deeper than Theorem 6.1. For example, while Theorem 6.1 is restricted to Markov processes for which the local time exists as a point indexed process, there are many interesting cases where the local time exists only as a distribution. This, of course, makes all the hard work we have done treating Gaussian processes on general parameter spaces even more justified.

Furthermore, by studying additive functionals of Markov processes it is possible to determine conditions for certain set-indexed Gaussian process themselves to have a kind of Markov property (Dynkin (1980)), and the entire $\mathcal{L}^2$ space of certain Gaussian processes defined on spaces of measures and certain generalised Gaussian processes defined on Schwartz space can be characterised via additive functionals of vector Markov processes (Adler and Epstein, (1986)). (This paper also attempts to explain "why" the isomorphism theorem works, and has a reasonable amount of expository material on the general background required for its appreciation.)

A proof of Theorem 6.1 (in a much wider setting) is given in Dynkin (1984a). Other related papers, in which you can find applications and extensions, are Dynkin (1983,1984b) and Adler, Marcus and Zinn (1989). All told, this is an exciting new area of activity, since it is one of the few areas in Probability Theory today that seems to be bringing people of diverse interests together, as opposed to what seems to be the natural trend for mathematicians to become more and more specialised and isolated even from their own colleagues. The reason you should be interested, is that the fact that you got this far into these notes indicates that you have a serious interest in Gaussian processes, and therefore may have something to contribute at this rapidly developing interface.

2. Directions for research.

MAJORISING MEASURES: As I mentioned a long time ago, in the Preface, my original motivation in preparing these Notes was to learn about majorising measures, which, at the time I started work, had only recently come to the fore in the literature of Gaussian processes. At the time, there was much excitement, for Talagrand's wielding of this concept had brought a solution to the fifty year old problem of finding necessary and sufficient conditions for sample path continuity.

Looking back over Talagrand's main result, however, it is clear that we have not progressed quite as much as we might like, for, at this point of time, we still do not understand majorising measures well enough to construct them in general. *If* an entropy condition is satisfied, whether it involves the simple entropy function of Dudley or the two-parameter entropy of Chapter 5, we can build a majorising measure. But, in these cases, we already knew how to handle continuity problems without majorising measures.

The challenge that therefore remains is to either more fully understand majorising measures, or to replace them with a concept that is more amenable to investigation. My personal feeling is that a two-parameter entropy approach will turn out to be the most natural in the long run, but I have been wrong more than once in the past (Wife – personal communication).

In view of Theorem 4.2, it also follows that majorising measures must have much in common with the distribution of the position of the maximum of X, when this is well defined. This is clearly an interesting problem in its own right, and very little is known. It is therefore doubly worthy of study.

EXTREMAL THEORY AND ASYMPTOTIC GROWTH PROPERTIES: To describe this problem, consider the following simple example: Let $X_t, t \in \Re$ be a centered, stationary, Gaussian process with continuously differentiable sample paths. Set

$$M(t) = \sup_{s \in [0,t]} X(s), \tag{6.5}$$

and, for $\lambda > 0$ and Borel $A \subset \Re$

$$N_A(\lambda) = \#\{t \in A \colon X_t = \lambda,\ X_t' > 0\}, \tag{6.6}$$

$$L_A(\lambda) = \nu(\{t \in A \colon X_t \geq \lambda\}), \tag{6.7}$$

where $\nu(B)$ is the Lebesgue measure of B. Then $N_A(\lambda)$ is the number of upcrossings of the level λ that X has in A, while $L_A(\lambda)$ is the amount of time, while in A, that X spends above the level λ.

There is an old and rich theory associated with the study of the three limits

$$\lim_{t,\lambda \to \infty} P\{M(t) > \lambda\},$$

as t and λ tend to infinity in a coordinated fashion, and

$$\lim_{x,\lambda \to \infty} N_{xA}(\lambda), \qquad \lim_{x,\lambda \to \infty} L_{xA}(\lambda),$$

where, again, x and λ tend to infinity in a coordinated fashion, $xA = \{xt : t \in A\}$, and the limit is to be taken in the sense of weak convergence of random measures.

You can find details of what happens in the limit (for the non-Gaussian, non-differentiable as well as Gaussian differentiable case) in Leadbetter, Lindgren and Rootzén (1983), Leadbetter and Rootzén (1988), and Leadbetter (1987), along with their copious bibliographies. In brief, if X satisfies an appropriate mixing condition, (i.e. $EX_tX_0 \to 0$ fast enough as $t \to \infty$), then the appropriate limiting distribution for M is double exponential, and for N and L the limiting processes are, respectively, Poisson and compound Poisson random measures.

All of these results, however, deal only with Gaussian processes on the real line. There is a rather immediate extension to processes on $\Re^d$, and you can find some older-style results in Adler (1981). There has never been an attempt, however, to extend the extremal theory of Gaussian processes to processes on general parameter spaces.

I leave it to you as a (heavily starred) exercise to see what can be made of this, and what applications would be generated by such results.

LOCAL, DIFFERENTIAL, STRUCTURE OF SAMPLE PATHS: There is a rich theory for Gaussian processes on Euclidean spaces that describes the structure of the sample paths in the vicinity of "rare events" such as local extrema at asymptotically high levels.

For example, let $X(t)$ be a continuously differentiable, stationary Gaussian process on $\Re^k$, with covariance function $R(t)$. Suppose that that at some point τ we have $X(\tau) = u$ and

$$\left.\frac{\partial X(t_1,\ldots,t_k)}{\partial t_i}\right|_{t=\tau} = d_i, \qquad i = 1,\ldots,k.$$

Then, with probability approaching one as $u \to \infty$, the random field has the following *deterministic* representation in a neighbourhood of τ:

$$(6.8)\quad X(t) = u + \sum_{i=1}^{k} d_i(t_i - \tau_i) + \tfrac{1}{2}u \sum_{i,j=1}^{k} (t_i - \tau_i)\lambda_{ij}(t_j - \tau_j) + O\left(\frac{1}{u}\right),$$

where

$$\lambda_{ij} = -\left.\frac{\partial^2 R(t)}{\partial t_i\,\partial t_j}\right|_{t=0}.$$

(For details on a precise formulation of this result see, for example, Leadbetter *et. al.* (1983) for processes on $\Re^1$ or Adler (1981) for random fields.)

Here is an interesting problem. We have developed throughout these notes most of the results that would be needed to extend results like the above to general parameter spaces. However, most parameter spaces will not have the nice differential structure that $\Re^k$ has, which is required to even write down (6.8). What is the extension of (6.8) to general parameter spaces, and how do you prove it? (I think the first question is the more difficult one.)

An answer to this question would, I believe, be of particular interest in the study of set indexed (i.e. empirical related) processes.

VI. REFERENCES

Adler, R.J. (1981), *The Geometry of Random Fields*, Wiley, Chichester.

Adler, R.J. (1984) The supremum of a particular Gaussian field, *Ann. Probability* **12** 436-444.

Adler, R.J. and Brown, L.D. (1986) Tail behaviour for suprema of empirical processes, *Ann. Probability* **14**, 1-30.

Adler, R.J., Brown, L.D. and Lu, K-L. (1990) Tests and confidence bands for bivariate cumulative distribution functions, *Communications in Statistics*, to appear.

Adler, R.J. and Epstein, R. (1987) Some central limit theorems for Markov paths and some properties of Gaussian random fields, *Stoch. Proc. Appls.* **24**, 157-202.

Adler, R.J. and Feigin, P.D. (1984) On the cadlaguity of random measures, *Ann. Probability* **12**, 615-630.

Adler, R.J., Marcus, M. B. and Zinn, J. (1990) Central limit theorems for the local times of certain Markov processes and the squares of Gaussian processes, *Ann. Prob.* **18** to appear.

Adler, R.J. and Samorodnitsky, G. (1987) Tail behaviour for the suprema of Gaussian processes with applications to empirical processes, *Ann. Probability* **15** 1339-1351.

Aldous, D. (1989), *Probability Approximations via the Poisson Clumping Heuristic*, Springer, New York.

Anderson, T.A., Giné, E., Ossiander, M. and Zinn. J. (1988) The central limit theorem and the law of the iterated logarithm for empirical processes under local conditions, *Probab. Th. Rel. Fields* **77** 271-305.

Bass, R.F. and Pyke, R. (1984) The existence of set indexed Lévy processes, *Z. Wahrsch. verw. Gebeite*, **66** 157-172.

Bass, R.F. and Pyke, R. (1985) The space $D(A)$ and weak convergence for set indexed processes, *Ann. Probability*, **13** 860-884.

Belyaev, Yu. K. (1961) Continuity and Hölder conditions for sample functions of stationary Gaussian processes, *Proc. Fourth Berkeley Symp. Math. Statist. Prob.* **2** 23-33, Univ. of California Press, Berkeley.

Berman, S. M. (1985) An asymptotic formula for the distribution of the maximum of a Gaussian process with stationary increments, *J. Appl. Prob.* **22** 454-460.

Berman, S. M. (1987) An extension of Plackett's differential equation for the multivariate normal density *SIAM J. Algebraic & Discrete Methods* **8** 196-197.

Borell, C. (1975) The Brunn-Minkowski inequality in Gauss space, *Invent. Math.* **30** 205-216.

Brown, L.D. and Rinott, Y. (1988) Inequalities for multivariate infinitely divisible processes, *Ann. Probability* **16**.

Cabaña, E.M. and Wschebor, M. (1982) The two parameter Brownian bridge – Kolmogorov inequalities and upper and lower bounds for the distribution of the maximum, *Ann. Probability* **10** 289-302.

Cambanis, S. and Rajput, B. M. (1973) Some zero-one laws for Gaussian processes, *Ann. Probability* **1** 304-312.

Cairoli, R. and Walsh J.(1975) Stochastic integrals in the plane, *Acta Math.* **134** 111-183.

Cramér, H. and Leadbetter, M.R. (1967), *Stationary and Related Stochastic Processes* Wiley, New York.

Dharamadhikari, S. W. and Joag-Dev, K. (1984) Regions whose probabilities increase with the correlation coefficient and Slepian's inequality, in Tong, Y. L. (1984) (ed.) *Inequalities in Statistics and Probability, IMS Lecture Notes-Monograph Series,* **5**, 156-164.

Dobric, V., Marcus, M. B. and Weber, M. (1988) The distribution of large values of a the supremum of a Gaussian process, *Asterisque* **157-8** 95-128.

Dobrushin, R.L. (1979) Gaussian and their subordinated self-similar random generalized fields, *Ann. Probability* **7** 1-28.

Dudley, R.M. (1967) The sizes of compact subsets of Hilbert space and continuity of Gaussian processes, *J. Funct. Anal.* **1** 290-330.

Dudley, R.M. (1973) Sample functions of the Gaussian process, *Ann. Probability* **1** 66-103.

Dudley, R.M. (1974) Metric entropy of some classes of sets with differentiable boundaries, *J. Approximation Theory* **10** 227-236.

Dudley, R.M. (1978a) Central limit theorems for empirical measures, *Ann. Probability* **6** 899-929.

Dudley, R.M. (1978b) Lower layers in $\Re^2$ and convex sets in $\Re^3$ are not GB classes, *Lecture Notes in Math* **709** 97-102.

Dudley, R.M. (1979) Lower layers in $\Re^2$ and convex sets in $\Re^3$ are not GB classes, *Lecture Notes in Math.* **709** 97-102, Springer Verlag.

Dudley, R.M. (1984) A course on empirical processes, (École d'Éte de Probabilités de Saint-Flour XII - 1982), *Lecture Notes in Math.* **1097** Springer Verlag.

Dvoretzky, A. (1961) Some results on convex bodies and Banach spaces, *Proc. Symp. on Linear Spaces*, 123-160, Jerusalem.

Dynkin, E.B. (1980) Markov processes and random fields, *Bull. Amer. Math. Soc.* **3** 957-1000.

Dynkin, E. B. (1983) Markov processes as a tool in field theory, *J. Functional Anal.*, **50** 167-187.

Dynkin, E.B. (1984a) Gaussian and non-Gaussian random fields associated with a Markov process, *J. Funct. Anal.* **55** 344-376.

Dynkin, E. B. (1984b) Polynomials of the occupation field and related random fields, *J. Functional Anal.*, **58** 20-52.

Dynkin, E. B. and Mandelbaum A. (1983) Symmetric statistics, Poisson point processes and multiple Wiener integrals, *Ann. Statist.*, **11** 739-745.

Ellis, R. (1988) Inequalities for multivariate compound Poisson distributions, *Ann. Probability* **16**

Evans, S.N. (1988a) Continuity properties of Gaussian stochastic processes indexed by a local field, *Proc. London Math. Soc.* **56** 380-416.

Evans, S.N. (1988b) Sample path properties of Gaussian stochastic processes indexed by a local field, *Proc. London Math. Soc.* **56** 580-624.

Evans, S.N. (1989) The maximum distribution of a Gaussian stochastic process indexed by a local field, *J. Austral. Math. Soc.* to appear.

Feldman, R. Epstein, (1989) Autoregressive processes and first-hit probabilities for randomized random walks, preprint.

Feldman, R. Epstein, (1990) Exit distributions for Lévy processes via Gaussian techniques, preprint.

Feller, W. (1971) *An Introduction to Probablity Theory and its Applications* **2** Wiley, New York.

Fernique, X. (1974) Minoration des functions aléatoires Gaussiennes, *Ann. Inst. Fourier* **24** 61-66.

Fernique, X. (1975) Regularité des trajectories des fontions aleatoires Gaussiennes, *Lecture Notes in Math.* **480** 1-96, Springer Verlag.

Fernique, X. (1976) Evaluations de processus Gaussiens composes, *Lecture Notes in Math.* **526** 64-84, Springer Verlag.

Fernique, X. (1978) Caractérisation des processus à trajectoires majorées ou continues, *Lecture Notes in Math.* **649** 691-706, Springer Verlag.

Fernique, X. (1983) Régularité de fonctions aléatoires non-Gaussiennes, *Lecture Notes in Math.* **976** 1-74, Springer Verlag.

Fernique, X. (1987a) Régularité de fonctions aléatoires Gaussiennes stationaires à valeurs vectorielles, *Probability Theory on Vector Spaces* to appear.

Fernique, X. (1987b) Fonctions aléatoires dans les espaces Lusiniens, *Espositiones Math.* to appear.

Fernique, X. (1988) Fonctions aléatoires à valeurs vectorielles, preprint.

Garsia, A.M., Rodemich, E. and Rumsey, H. (1970) A real variable lemma and the continuity of paths of some Gaussian processes, *Indiana Univ. Math. J.* **20** 565-578.

Geman, D. (1977) On the approximate local growth of multidimensional random fields, *Z. Wahrsch. verw. Gebeite* **38** 237-251.

Geman, D. (1978) Dispersion points for linear sets and approximate moduli for some stochastic processes, *Trans. Amer. Math. Soc.* **253** 257-272.

Geman, D. (198?) The approximate growth of Gaussian processes, *Unpublished preprint.*

Geman, D. and Zinn, J. (1978) On the increments of multidimensional random fields, *Ann. Probability* **6** 151-158.

Giné, E., Hahn, M.C. and Zinn, J. (1983) Limit theorems for random sets: An application of probability in Banach space results, *Lecture Notes in Math.* **990** 112-135, Springer Verlag.

Goodman, V. (1975) Distribution estimates for functionals of the two-parameter Wiener process, *Ann. Probability* **4** 977-982.

Gordon, Y. (1985) Some inequalities for Gaussian processes and applications, *Israel J. Math.* **50** 265-289.

Gordon, Y. (1987) Elliptically contoured distributions, *Prob. Theory Rel. Fields* **76** 429-438.

Gordon, Y. (1988a) Gaussian processes and almost spherical sections of convex bodies, *Ann. Probability* **16** 180-188.

Gordon, Y. (1988b) On Milman's inequality and random subspaces which escape through a mesh in $\Re^n$, *Lecture Notes in Math.* **1317** 265-289, Springer Verlag.

Hogan, M. L. and Siegmund, D. (1986) Large deviations for the maxima of some random fields, *Adv. Appl. Math.* **7** 2-22.

Iscoe, I., Marcus, M.B., McDonald, D., Talagrand, M. and Zinn, J. (1990) Continuity of ℓ^2-valued Ornstein-Uhlenbeck processes, *Ann. Probability* **18** to appear.

Itô, K. (1983) Distribution valued processes arising from independent Brownian motions, *Math. Z.* **182** 17-33.

Itô, K. and Nisio, M. (1968) On the convergence of sums of independent Banach space valued random variables, *Osaka J. Math.* **5** 35-48.

Jain, N.C. (1971), A zero-one law for Gaussian processes, *Proc. Amer. Math. Soc.* **29** 585-587.

Jain, N.C. and Marcus, M.B. (1978) Continuity of sub-Gaussian processes, *Advances in Probability* **4**, M. Dekker, New York.

Joag-Dev, K., Perlman, M.D. and Pitt, L.D. (1983), Association of normal random variables and Slepian's inequality, *Ann. Probability* **11** 451-455.

Kac, M. Kiefer, J. and Wolfowitz, J.(1955) On tests of normality and other tests of goodness of fit based on distance methods, *Ann. Math. Statist.* **26** 189-211.

Kahane, J-P. (1986) Une inegalite du type de Slepian et Gordon sur les processus Gaussiens, *Israel J. Math.* **55** 109-110.

Kallianpur, G. (1969), *Lecture Notes*, Indian Statistical Institute, Calcutta

Kallianpur, G. (1970), Zero-one laws for Gaussian processes, *Trans. Amer. Math. Soc.* **149** 199-211.

Karatzas, I. and Shreve, S.E. (1988) *Brownian Motion and Stochastic Calculus* Springer Verlag, New York.

Kolmogorov, A. N. and Tihomirov, V. M. (1959) ϵ-entropy and ϵ-capacity of sets in functional spaces, (in Russian) *Usp. Mat. Nauk.* **14** 1-86. [English translation, (1961) *Am. Math. Soc. Transl.* **17** 277-364.]

Landau, H. and Shepp, L.A. (1970) On the supremum of a Gaussian process, *Sankya* **32** 369-378.

Leadbetter, M.R. (1987) On the exceedence random measure for a stationary process, Preprint.

Leadbetter, M.R., Lindgren, G. and Rootzén, H. (1983) *Extremes and Related Properties of Random Sequences and Processes* Springer Verlag, New York.

Leadbetter, M.R. and Rootzén, H. (1988) Extremal theory for stochastic processes, *Ann. Probability* **16** 431-478.

Ledoux, M. (1988) Inégalités isopérimétriques et calcul stochastique, *Lecture Notes in Math.* **1321** 249-259, Springer Verlag.

Linde, W. (1986) *Probability in Banach Spaces – Stable and Infinitely Divisible Distributions*, Wiley, Chichester.

Major, P. (1981) *Multiple Wiener-Itô Integrals, Lecture Notes in Math.* **849**, Springer-Verlag.

Marcus, M. B. (1987) ξ-radial processes and random Fourier series, *Memoirs Amer. Math. Soc.* **368** 1-181.

Marcus, M.B. and Pisier, G. (1981) *Random Fourier Series with Applications to Harmonic Analysis* Annals of Math. Studies 101, Princeton Univ. Press, Princeton.

Marcus, M.B. and Pisier, G. (1984) Characterizations of almost surely continuous p-stable random Fourier series and strongly stationary processes, *Acta Math.* **152** 245-301.

Marcus, M.B. and Rosen, J. (1990) Sample path properties of the local times of symmetric Markov processes, In preparation.

Marcus, M.B. and Shepp, L.A. (1971) Sample behaviour of Gaussian processes, *Proc. Sixth Berkeley Symp. Math. Statist. Prob.* **2** 423-442.

Milman, V.D. and Schechtman, G. (1986) Asymptotic Theory of Finite Dimensional Normed Spaces, *Lecture Notes in Math.* **1200** 265-289, Springer Verlag.

Orsingher, E. (1987) On the maximum of random fields represented by stochastic integrals over circles, *J. Appl. Prob.* **24** 574-585.

Orsingher, E. and Bassan, B. (1988) On the comparison of the distribution of the maximum of random fields represented by stochastic integrals over circles, preprint.

Pickands, J. III (1969a) Upcrossing probabilities for stationary Gaussian processes, *Trans. Amer. Math. Soc.* **145** 51-73.

Pickands, J. III (1969b) Asymptotic properties of the maximum in a stationary Gaussian process, *Trans. Amer. Math. Soc.* **145** 75-86.

Pisier, G. (1980) Conditions d'entropie assurant la continuité des certains processus at applications à l'analyse harmonique, *Séminaire d'Analyse Fonctionnelle 1979/80*, Ecole Polytechnique, Paris.

Pisier, G. (1986) Some applications of metric entropy to harmonic analysis, *Lecture Notes in Math.* **995** 123-154, Springer Verlag.

Pisier, G. (1986) Probabilistic methods in the geometry of Banach space, *Lecture Notes in Math.* **1206** 167-241, Springer Verlag.

Pisier, G. (1989) *The Volume of Convex Bodies and Banach Space Geometry*, Cambridge Univ. Press, Cambridge.

Piterbarg, V.I. (1986) *Lectures in the Theory of Gaussian Processes*, (In Russian), Moscow Univ. Press.

Piterbarg, V.I. (1988) *Asymptotic Methods in the Theory of Gaussian Processes and Fields*, (In Russian), Moscow Univ. Press.

Plackett, R.L. (1954) A reduction formula for normal multivariate integrals, *Biometrika* **41** 351-360.

Preston, C. (1972) Continuity properties of some Gaussian processes, *Ann. Math. Statis.* **43** 285-292.

Pollard, D. (1984) *Convergence of Stochastic Processes*, Springer Verlag, New York,

Pollard, D. (1990) Cube root asymptotics, preprint.

Pyke, R. (1984) Asymptotic results for empirical and partial sum processes: A review, *Can. J. Statist.*, **12** 241-264.

Pyke, R. (1985) Opportunites for set-indexed empirical and quantile processes in inference, *Bull. International Statist. Inst.*, **51** *25.2* 1-11.

Riesz, R. and Sz-Nagy, B. (1955) *Functional Analysis*, Ungar, New York.

Samorodnitsky, G. (1987a) Bounds on the supremum distribution of Gaussian processes – Polynomial entropy case. Preprint.

Samorodnitsky, G. (1987b) Bounds on the supremum distribution of Gaussian processes – exponential entropy case. Preprint.

Samorodnitsky, G. (1988) Continuity of Gaussian processes, *Ann. Probability* **16**, 1019-1033.

Samorodnitsky, G. (1990) Probability tails of Gaussian suprema, *Stoch. Proc. Appls.* to appear.

Samorodnitsky, G. and Taqqu, M. (199?) *Stable Random Processes*, In preparation.

Siegmund, D. (1988) Approximate tail probabilities for the maxima of some random fields, *Ann. Probability* **16**, 487-501.

Slepian, D. (1961) First passage time for a particular Gaussian process, *Ann. Math. Statist.* **32** 610-612.

Slepian, D. (1962) The one-sided barrier problem for Gaussian noise, *Bell System Tech. J.* **42** 463-501.

Sudakov, N.V. (1969) Gaussian and Cauchy measures and ϵ-entropy, *Soviet Math. Dokl.* **10** 310-313.

Sudakov, N.V. (1971) Gaussian random processes, and measures of solid angles in Hilbert space, *Soviet Math. Dokl.* **12** 412-415.

Talagrand, M. (1984) Sur l'intégrabilité des vecteurs guassiens, *Z. Wahrsch. verw. Gebeite* **68** 1-8.

Talagrand, M. (1987) Regularity of Gaussian processes, *Acta Math.* **159** 99-149.

Talagrand, M. (1988a) Small tails for the supremum of a Gaussian process, *Ann. Inst. Henri Poincare* **24** 307-315.

Talagrand, M. (1988b) Necessary conditions for the sample boundedness of p-stable processes, *Ann. Probability* **16** 1584-1595.

Talagrand, M. (1989) On subsets of $\mathcal{L}^p$ and p-stable processes, *Ann. Inst. Henri POincaré* **25** 153-166.

Tong, Y. L. (1984) (ed.) *Inequalities in Statistics and Probability, IMS Lecture Notes-Monograph Series,* **5**.

Vapnik, V.N. and Červonenkis, A.Ya. (1971) On the uniform convergence of relative frequencies of events to their probabilities, *Theor. Probability. Appl.* **16** 264-280. Springer Verlag.

Vapnik, V.N. and Červonenkis, A.Ya. (1974) *Theory of Pattern Recognition: Statistical Problems in Learning* [In Russian] Nauka, Moskow.

Walsh, J.B. (1986) An introduction to stochastic partial differential equations, *Lecture Notes in Math. 1180* 265-437, Springer Verlag.

Weber, M. (1988) The supremum of Gaussian processes with a constant variance. Preprint.

Weron, A. (1984) Stable processes and measures: A survey, *Lecture Notes in Math.* **1080** 306-364, Springer Verlag.

Watanabe, S. (1984) *Stochastic Differential Equations and Malliavan Calculus*, Tata Inst. of Fundamental Research, Springer, Berlin.

Wong, E. and Zakai, M. (1976) Weak martingales and stochastic integrals in the plane, *Ann. Probability* **4** 570-586.

Zaanen, A.C. (1956) *Linear Analysis*, North Holland, Amsterdam.

Note: Major references to particular items are given in italics.